The Flemish System of Poultry Raising

by Madame Jasper

with an introduction by Jackson Chambers

Self Reliance Books

Get more historic titles on animal and stock breeding, gardening and old
fashioned skills by visiting us at:

http://selfreliancebooks.blogspot.com/

Introduction

I am pleased to present yet another title on Poultry.

The work is in the Public Domain and is re-printed here in accordance with Federal Laws.

As with all reprinted books of this age that are intended to perfectly reproduce the original edition, considerable pains and effort had to be undertaken to correct fading and sometimes outright damage to existing proofs of this title. At times, this task is quite monumental, requiring an almost total "rebuilding" of some pages from digital proofs of multiple copies. Despite this, imperfections still sometimes exist in the final proof and may detract from the visual appearance of the text.

I hope you enjoy reading this book as much as I enjoyed making it available to readers again.

Jackson Chambers

MADAME JASPER.

Dédié

A la Comtesse de Derby

En témoignage affectueux et respectueux de l'auteur

CONTENTS

LIST OF ILLUSTRATIONS

INTRODUCTORY NOTE

BEFORE the war there was no country in the world
in which the wholesale production of table poultry
was conducted on so great a scale as in Belgium,
especially in the neighbourhood of Brussels. Poultry keeping there is reduced almost to the certainty
of a factory. The skill with which this is done is
due in large measure to the age of the industry.
It has been carried on for centuries at Merchtem,
Londerzeel, Cappelle-au-bois, Malines, Heyst-op-
den-Berg, etc. The nickname of "chicken-eaters"
given to the people of Brussels in the Middle Ages
would itself go far to prove the antiquity of the
industry. Chickens in this neighbourhood are bred
by thousands and sold by thousands daily.

It seemed, then, an opportunity not to be missed,
when the fortune of war caused Madame Jasper to
seek a temporary refuge in this country, to induce
her to write this book. At her home, not far from
Tongres, she had, when hostilities were declared, a
very large establishment for chicken-rearing, and her
name was very widely known as that of one of the
most capable, inventive, and scientifically accom-
plished authorities on all that pertains to the breeding
and management of chickens. Originally she had
been attracted by the interest of the art and her

Introductory Note

extraordinary ingenuity led her to try many experiments as to the quickest, surest, and most efficacious methods. But a pleasant little hobby developed into a very large business, and her name became known far outside the limits of her own country.

In part this was due to the thoroughness with which she investigated the whole subject. Before attempting to rear chickens on a large scale, she visited every country in the world that had any reputation for producing chickens, Great Britain among the rest. In this country Madame Jasper found, as she has explained in this book, that our people are unrivalled in the skill with which they breed and prepare birds for exhibition, and there are no better breeds in the world than we have here, but no one had thoroughly systemised the art of production on a really large scale, and with an eye to make profits of a magnitude not to be scorned in any industry.

It will be found by those who peruse these pages that there is very little in the art of keeping poultry for which Madame Jasper has not suggested an improvement. The equipment she uses is cleverly directed either to the saving of labour or the improvement of results. Her book will interest all poultry keepers, and gives a wealth of useful hints and directions.

P. A. G.

THE FLEMISH SYSTEM OF POULTRY REARING

CHAPTER I

INTRODUCTION

BEFORE beginning this series of articles on the Flemish rearing of table poultry, I wish to express my indebtedness to English methods. From them I first derived my taste for poultry keeping ; it was they who made me a lady farmer, and if I have attained to any fame in this department, I owe it partly to England. It is all the more important for me to begin by expressing my admiration for, and my gratitude to, England, since I am about to criticise English methods. For I intend to prove that if the English are our masters in the art of breeding, they have, on the whole, hitherto shown themselves ignorant of those methods for producing poultry for the market which bring in the most profit.

Lectures and Articles.

When I first appeared as a writer on this subject in the Brussels paper *Chasse et Pêche*, about twelve years ago, I translated articles from *Poultry*,

Introduction

Feathered World, *Poultry for the Many*, and I quoted from the work of Tegetmeier, Lewis Wright, Harrison Weir, etc., until, as the result of several years' experience and observation, I was able to form opinions of my own arising from facts which I had carefully studied and mastered. Henceforth my articles became the subject of bitter controversy; they were approved by some, condemned by others. How could I dare to criticise things which until then had been considered as articles of faith? How could I venture to form my own opinions, to deny established facts? What arrogance, and what a stir among the sheep of Panurge!

After a while, however, controversy died down. So many were the visits and the letters of those who daily wished to consult me that I was compelled to refuse to reply to them. My articles and my lectures were reproduced in France, in *La Vie à la Campagne*, in several of Hatchette's publications, in agricultural magazines, etc. One of my last articles in dialogue (illustrated by Léon Defrecheux, 1910–1911), " L'Arrivée d'une jeune Poulette à Mon Plaisir," was considered so amusing and instructive that it was reproduced as a serial story in *La Meuse*, a Liége paper with a very large circulation. My hens, all Buff Orpingtons, became heroines; they were known by the name of " Belle-Orpigne," " Hen 308," " Poulette," etc. So great was the fame they brought me that during a motor tour in France and Belgium I found myself called " Madame Belle-Orpigne." A few months before the war I was asked by M. Hyppolite Rollin, the Secretary of the Belgian Minister of Agriculture, whether I would consent to the publication of my articles in an illustrated volume. Accordingly, I asked the papers of *La Meuse* and *Chasse et Pêche* to return my negatives, and in July, 1914, I was preparing this book, when war broke out. Since then I have made several efforts to obtain my articles and negatives, but without success, alas! owing to military censorship and to the German occupation of Belgium.

Before establishing my industry commercially, which I was doing when war drove me from my home, for nine years I had occupied myself in poultry keeping, merely as a hobby, taking a delight in sending eggs and hens as presents to my friends, to hospitals, and to the needy poor. I had been spending on various experiments more than £200 a year; I had been raising annually 1,500 to 2,000 hens, experimenting, incubating, selecting and mating my breeding and laying stock according to English methods. Some of my birds, my incubators, foster mothers, etc., were even bought in this country. I myself personally carried out—and this is very important—the most insignificant details, for without this personal work such an occupation cannot possibly be successful. " No one can be a good master without first being a servant " is a saying which is especially true for a trade which consists in minutiæ, constant care, personal supervision and observation.

The Science of Poultry Breeding.

In a short time the intensity of its interest and of its science rendered poultry keeping my dominant

Introduction

passion, my constant preoccupation and the delight of my life. What happy memories, what charming moments, I owe to this study! For my broods I became a tender mother, a skilled and thoughtful nurse. To these characteristics I added a patience and a desire for knowledge fostered by my reading of critical works on this subject in French and in German. I plunged into the most serious studies of the science of breeding in general : into the chemical questions of feeding, into the physiological questions of digestion, productibility, fixation of food, etc.

For some years 25,000 to 30,000 chickens bred for the market have every year passed before me, thus widening my field of experience, and to-day rendering me capable of discussing these questions with intelligent experts in this country.

I had formed the idea of instituting in England a model poultry farm for the intensive and artificial production all the year round of chickens for the table, like the establishments which we already have in Belgium, and especially like that on my own estate. I should like to find in this country people anxious to give their names to such a work, which would bestow upon England a new and profitable industry. For a large number it might constitute a source of wealth and prosperity. The demonstrations that might be carried on there, the practical experiments conducted, would have an educative influence far wider, more durable, and more striking than all the articles written on the subject. Perhaps the present time is hardly favourable for such an undertaking ; nevertheless, I do not despair of one day realising my dream.

English Breeders and Fatteners

The famous English Breeds.

A sudden change in my life induced me some years ago to invest part of my capital in the establishment of a poultry farm on commercial lines. Into this farm I put all my knowledge and all my taste, devoting to this industry the most modern discoveries and endeavouring at the same time to establish it on a profit-making basis. I had first travelled in Germany, in France and in Holland, but I had not visited England. I knew that the English were excellent in the production of new breeds, sporting breeds and fancy fowls. I knew that hitherto as breeders they were unequalled, and that to them we owe the superb breeds of Durhams, Shorthorns, Southdowns, Yorkshires, etc., which command the admiration of the whole world. I knew also that they had produced first-rate breeds of poultry. I knew that their crossings were judicious and excellent. But I knew also that to them the art of the rapid production of a high average quality of table chickens was almost entirely unknown. That does not mean that in England there are not breeders and fatteners who are very competent; but there are far too many who produce absolutely inferior poultry, and it is they especially who need instruction in new methods. I knew that the English, because of their love of the open air, knew nothing of the art of raising poultry, artificially produced according to our methods. Those methods are repulsive to their sporting tastes, their love of the open air and of muscular exercises. But such proclivities are not

5

Introduction

favourable to the production of a rapidly grown, artificially produced chicken, with young, white, soft and savoury meat.

Returning from my travels, it was borne in upon me that, in accordance with the proverb, "A prophet is not without honour save in his own country," I had been to great expense to seek in distant lands something which lay at my very door. For I discovered that there is no one like the Flemings in this art of rapid and remunerative production. If it was not absolutely superior in quality to the French and English productions of table fowl, at least it was equal to anything I had seen. From generation to generation our peasants have been engaged in poultry keeping.

Poultry Keeping in Belgium.

History teaches us that as early as the ninth century the Flemings were renowned for their skill in producing chickens for the table, and in the whole art of poultry keeping. I knew that since that time poultry keeping had been so far developed in Belgium that whole districts were employed entirely in the production. The Belgians have not, like the English, produced a great variety of breeds, but their breeds, such as Brackel, Cambine, Herve, Malines, are perfectly adapted to their environment. I knew that the qualities of these breeds had been thoroughly developed, and that the Belgians excelled in the fattening and the production of table chickens. I was also aware that the fertility of the country was largely due to the manuring of the ground by the poultry kept upon it.

Gathering Experience

But I also knew what errors they had made, and how ignorant they were of science. I believed that as much, or even more, could be done in other countries. Only intelligent comparison could enlighten me on this point. My expensive journeys, therefore, had not been useless; they showed me that the opinions held by the ancients as to our skill in poultry keeping were completely justified.

I had seen what was done elsewhere, the good and the evil of it; and in my own poultry farm I have avoided committing like errors. In every science, and in its application, there is something to be taken and something to be left. Then I added a few improvements—details of my own invention, and some which I had observed during my travels.

Since October, 1914, the date of my arrival in this country, owing to the kindness of the Minister of Agriculture, I have been placed in a position thoroughly to investigate the question of the production of table poultry and production of eggs. My numerous visits to different poultry farms have only confirmed me in my belief that, except in the case of certain rare and clever producers, English poultry is of an inferior quality. I was further confirmed in my opinion by my visit to the Central Market, where Mr. Charles E. Brooke, a charming gentleman, a very competent breeder and a great poultry dealer, acted as my guide. The numerous inferior birds offered for sale showed me that the science is not yet widely understood in England. Though there were a few fine specimens, how many were there of a totally inferior quality !

For some years the introduction of certain Ameri-

Introduction

can methods modified by common-sense, necessity and practice, have made a commercial industry of poultry keeping in my country, and nowhere is it so remunerative as in Belgium. The installation of great poultry farms dates from this period. Our gentlemen farmers, investing their capital in this industry, have begun to produce from 10,000 to 50,000 birds a year, partly table birds and partly chickens sold for breeding.

Prosperous Belgian Poultry Keepers.

The neighbourhoods of Brussels, Antwerp and Malines are the great centres for the production of eggs, chickens and fowls. The production and the fattening of the birds has here attained something like perfection. Does that mean that everything is excellent? Certainly not. There is still better to be done. Our peasants are not scientific, and the splendid English methods of breeding stock are too often unknown to them. Volumes might be written on the stupidity, the inconsequence and the super-annuated methods of our country folk. In the course of these studies I shall have occasion to return to this subject. Suffice it now to say that the industry has been one of the sources of our national prosperity.

The population of Belgium to the square mile is greater than that of any other country in the world. Our population not only feeds itself, but exports considerable quantities of eggs and poultry. In 1905 we are said to have exported £2,000,000 worth. Since then our exports have certainly increased. The difficulty of consulting documents containing

Educating the Peasants

official statistics makes it impossible for me to give more recent figures. All I can say is that we annually export millions of dead birds and that our productions are universally held, not only in France and Germany, but in this country also, to be of superior quality. This intensive production and the spur given to this industry by the profits derived from it have attracted the attention of the public authorities. For some years our National Federation of Aviculture, financed by the Belgian Government, has undertaken to educate our peasants in this science, in which, curiously enough, they had made but little progress for centuries. Hence the necessity of teaching them better methods of hygiene and of breeding, in order to develop still further this source of prosperity for our industrious country.

The Objects in View.

In short, the object of this science may be expressed in a few words: rapid and intensive production, at a slight cost, of a bird ready for consumption, and, in the case of large poultry farms, we may add, the highest interest possible on the invested capital.

And in a more elevated, more humanitarian domain we believe that an establishment conducted on strictly commercial lines will also be educative, that it will open up new horizons, that it will create new needs, new branches of activity; that it will, therefore, provide employment which does not call for great strength or energy, but demands only care and attention to the minutest detail. It will then

Introduction

afford a most providential field of activity for poor cripples, victims of this cruel war ; it will create a new life and contribute to social well-being by raising up new centres of population in deserted spots ; it will also contribute to the national funds, and it will help England to be self-sufficing by producing her own eggs and birds for her own consumption. This, in every country, is of the greatest importance.

CHAPTER II

FLEMISH AND ENGLISH METHODS

POULTRY keeping is the mother of many and various industries; for example, certain Belgian dealers collect eggs from a number of neighbouring farms, and furnish their customers with 50,000 to 60,000 a month, sending at once sometimes 6,000, 8,000, 15,000 and even more. These dealers are numerous, and in the collection of the eggs they generally employ old people and children incapable of engaging in other occupations. In order to obtain eggs in such great quantities, large numbers of laying hens are necessary. All the villagers in the neighbourhoods of Merchtem, Londerzeel, Malines, Aerschot, Heyst-Op-Den-Berg, Puers, etc., keep poultry of the breeds belonging to the country, and demanded by the market. Every week hundreds of thousands of eggs are sent out of these villages to poultry keepers, who incubate chickens and keep them until they are fit for sale; that is to say, about twelve or fourteen weeks, according to the nature of the farm, the requirements of the market and the system of feeding.

Boarding out Fowls with Peasants.

Certain large poultry farms, with the object of obtaining absolutely new-laid eggs, laid by birds

Flemish and English Methods

which they know to be thoroughly healthy, board out cocks and hens with the peasants. These peasants are forbidden to keep other breeds; they must either completely or partially feed the birds according to the agreement, and are obliged to sell to the owner all the eggs produced by them, generally at a price slightly higher than that paid at the market. Thus they have no interest in cheating and selling the eggs elsewhere. The eggs are collected several times a week. The owners of the birds reserve to themselves the right to take them away from the houses where they are badly fed, not properly kept, or otherwise neglected. The stock is removed every two years; the old hens are sent to the market, and it is hardly necessary to say that both producers and reproducers are kept freely without restraint till the moment they are required to lay.

The hatching time is regulated by the time at which the birds are desired to lay. The ideal plan would be personally to superintend the production of the eggs if the promiscuity of hens and chickens did not render possible and dangerous the propagation of disease and of vermin, so disastrous to the development of young birds. A verminous fowl is a bad subject for fattening. Not being able to rest on account of the continuous irritation to which it is exposed, it does not put on flesh so rapidly, and the actual loss of chicken meat through the ravages of vermin must form a considerable item in any large establishment. Scientific poultry keepers do not generally keep their own breeding stock. The difficulty often experienced in obtaining sufficient ground, and the continual drawback, that it necessitates a double staff which does not come into contact at all with the breeding stock, has prevented them from doing so. Later I intend to solve this question and to show how in the future these dangers may be avoided and laying stock may be kept under one's own personal control, thus avoiding the double danger of unscrupulous dealers and of careless peasants.

Other poultry people, who go by the name of *accouveurs*, have sometimes superb establishments, in which forty or fifty incubators of a capacity of 400 and 500 eggs, are constantly at work. Every week, or even every day, thousands of hatched chickens are sent to poultry keepers, who prefer not to incubate at home.

" Coarse, Yellowish Fowls."

Others, again, buy from poultry keepers on a small or a large scale chickens which they sell, as they are bought either as corn-fed chickens, or as fine fat birds for the table. Neither in Belgium nor in France are ever to be seen those coarse, yellowish fowls with gross hard skin, five months old or more, that are too common in England. We would scruple to use them as boiling fowls, and certainly would never dare to serve them as roast chicken on any decent table. We have long discarded all yellowish or violet meat still to be seen in this country. We do not produce for table birds older than three months. There used to be one exception; we formerly bred a big bird which was greatly in demand in Germany, but we reared and fed it

Flemish and English Methods

according to a very special system until it had attained the age of five months; this bird having retained its tenderness. In Belgium we take great pride on having young birds with fine, snow white meat, with pink and flexible claws. All those who deal in this admirable poultry, not sufficiently appreciated in England, can make a good living out of the industry. The prosperity of our country folk bears eloquent testimony to this fact.

Nevertheless our peasants and breeders have much to learn; they can do still better, and I hope that in the future they will appreciate more the enormous extent to which the industry of poultry rearing can be developed. Neither in England nor in Belgium are the birds made to yield the best possible results. Of course I am speaking of the generality. During my recent stay in England I have visited certain modern English poultry farms which are admirably conducted, very much on the same lines as the best establishments in my own country. In Belgium, the weakness lies chiefly in the lack of an intelligent understanding of three great questions, namely, (1) those hygienic laws which must be obeyed to obtain perfect health in the chickens, (2) the evil effect of inbreeding, (3) the science of selecting. In England, the trouble originates in a different source. English people have a double ideal in their minds in regard to poultry breeding. They have one eye on exhibition birds and the other on utility fowls. These objects cannot be pursued at one and the same time. A prize bird must lay, of course—it may even be a good layer—but rarely will it attain its maximum.

Tegetmeier's Opinion

Here, as in everything else, specialisation makes perfection. A good layer will probably be ugly, thin and active, which is not the case with a prize bird, which must correspond, above all things, to a certain standard, and may or may not be a good layer.

Harm wrought by Poultry Fanciers.

The fancy fowl has done no end of harm to the production of eggs and good meat. In this connection let me quote the opinion of Tegetmeier, who, in my opinion, was not sufficiently appreciated by his own countrymen. " I do not hesitate to affirm, as the result of my experience of half a century, that no breed of fowls has been taken in hand by the fancier that has not been seriously depreciated as a useful variety of poultry. At shows, as at present conducted, fancy points only have to be considered by the judges; the result is that the economic value of many breeds has been entirely lost." Nothing is more true. And how many English breeds, admirable both for eggs and for the table, have been irremediably spoilt by the standard which demands a certain type of conventional beauty without considering other qualities ?

From the commercial point of view, nothing can be better than the common fowl accustomed to be exposed to wind and weather of all sorts, nothing can be better if they were selected than mongrels, which type no one has ever attempted to improve. Nothing lays better than the native fowl, for it is acclimatised to the atmosphere in which it lives, to the ground on which it runs, to the grass which

Flemish and English Methods

grows there, and, in my opinion, it is always a great mistake to take birds from one country to another. The best layers, therefore, are the hens of the country, *selected* with this object. This in parentheses. I know that in England there are competitions of laying birds, specially kept and fed for high production. These have all our admiration, and we do not expect beauty from them. We shall choose our birds for their productiveness, for their excellent meat and for their rapid development, and by judicious selection and careful feeding we shall try to convert them into good layers. I know how to discover these productive faculties without having recourse to trap nests. We shall attain it under certain conditions, for our industry depends on a regular production of eggs and we must endeavour to avoid any cessation of laying. I should say that the first step in poultry keeping, according to Flemish methods, should be the application of the good and healthy English system of poultry keeping. Our layers might run and develop in the open air when they are young and we should avoid too much consanguinity, good for excellence of type, but injurious as tending to degeneracy and other consequences. In the course of this study I will endeavour to show the various purposes to which eggs can be applied and, later on, how large quantities of good and fertile eggs may be obtained. As soon as this industry is established in England as it was in Belgium before the outbreak of war, attention will be directed to the more numerous and interesting branches of this trade. It may be expected that *accouveurs* will come into existence in far greater numbers than they ever have been in England before. They will devote themselves exclusively to the production of breeds of chickens suitable for the table. Subsidiary branches of the industry will be the rearing of little milk chickens, little Royal chickens, Asparagus chickens and little Cocotte or Hamburg chickens, the last mentioned being produced by a very interesting crossing of which particulars will be given later on. In this country little chickens of this kind are very few, because of the scarcity of cockerels devoted specially to this use. In Belgium it is a special production including cockerels and pullets. These small birds, ready for sale after five or six weeks, are greatly in demand in the market.

We should also produce young poultry called spring chickens, corn-fed chickens, and we should develop the fattening industry. On this point also I shall have much to say, because fattening, as it is understood in this country, shows an enormous wastage of food, and produces a result which, if not inferior, is rarely superior in quality to that which we obtain by our Flemish method. In certain poulterers' shops I have seen very fine poultry, but on account of the irrational method of keeping it it does not bring the same profit.

We shall begin our first incubation in October, and our first sales towards the end of December or the beginning of January. In January and February prices are high and profits are extremely good. In England most of the new chickens do not reach the market before March ; in Belgium we have them all the year round, and so great is the demand that we can never satisfy it. We have very little to fear in

Flemish and English Methods

the way of competition. One great poultry dealer in this country said to me the other day, "Anyone who can produce a fine table chicken all the year round will command not only the market but the price."

Raising Chickens in Winter.

Naturally, the greatest profits are made when dealings are conducted on a large scale, but in any case, large or small, the industry should be very profitable. The buildings employed in our industry permit of the production of chickens in the early months of winter, and here we find the force of our methods. Did one ever succeed in producing early fruits and vegetables, strawberries, asparagus, by exposing them to the cold, to the winds and winter damp, or to other atmospheric changes? Certainly not. A chicken raised in winter is exactly like these other early productions. I have seen in this country on poultry farms which I visited in December, January and February, poor little chickens shivering on the damp grass, their down soaked, and their tender organisms at the mercy of the bitter wind. How is it possible to obtain any rapid growth under these conditions? And think of the cost of the superfluous food with which you are providing them merely to produce the necessary calories which shall help them to struggle against the bitterness of cold and damp. All this food, in our method, goes to the production of flesh, and is not uselessly wasted. And what a superfluous expenditure of labour is required to look after all those little rearing houses scattered over the grass, work which might be so much better applied in cleansing, in the filling of lamps and in the giving of food. To say nothing of these cramped little houses where the tiny creatures sleep in an atmosphere lacking air and too rapidly vitiated. There is nothing rational in the manner in which poultry is reared in England for table production. How ridiculous and completely unproductive are such methods I intend to demonstrate later on.

Protecting Chickens from Cold.

The distinguishing feature of our Flemish system is that of sheltering the birds from cold and damp, instead of depending upon food to counteract the effect of these agencies. At the same time, in their clean, well-lighted houses our birds have plenty of air without any draught, a mild and carefully regulated temperature, and a clean and spacious brooder in which they are put to sleep. Their houses are cleaned every day and not merely once or twice a week. Thus it is impossible for them to have any vermin, which cannot be avoided when night after night they not only sleep on filth, but in a vitiated atmosphere which injures their health and hinders their growth. They are never allowed to run on grass already contaminated by hens from which not only vermin may be caught, but illness may be contracted.

We shall study the various details of this method, details of commercial incubation, which must be carried on in a somewhat different manner if 8,000 or 10,000 eggs are to be produced in a month, and we shall survey all the details connected with this very interesting occupation.

CHAPTER III

INCUBATORS

In order to rear economically good table fowls which would leave a substantial profit, we must first learn how to obtain the greatest percentage out of the eggs we incubate.

Some fifteen years ago our Flemish peasants used to incubate with hens ; now very few of them do so. In the smallest farms where production of table birds is carried out we no longer see broody hens. Very quick where his interest is concerned, the Flemish peasant promptly realised that with a hen, chickens could not be produced in the season of the year when table fowls, being scarce, fetch high prices. The hen cannot be made to brood at will, and the custom of setting hen turkeys on eggs when it is suitable, as is done in some parts of France, is not in favour in Flanders. So the production of table fowls, carried out on a small or big scale, should be effected by means of incubators. With them there is no breakage of eggs if gently handled ; no crushing of the chick by a stupid mother, nor eating of the eggs by a voracious one, and *no vermin*—a cause of so many failures where rapid production is sought for. With incubators, a great number of chicks can be produced at will, and at the time of year when

The Advantage of Numbers

the market requires them. In my establishment I incubate only from October till May. Some people continue all the year round, but birds are not so highly profitable in summer as in winter, but this is done with the object of keeping the attendants occupied and so covering the running expenses of the undertaking and extracting all possible profit from it.

Hatching in Large Numbers.

The greatest profits will be obtained where the greatest percentage of chicks is hatched. In our big, light and well-aired brooder houses accommodation is made for a great number of chicks—500 or 1,000, more or less, will make no difference in light, heating or time ; even food will increase the expenses very little. We must bear in mind that every egg lost, every chick dead in shell, every cripple, raises the average prices of the others. Consequently, it is most important and more than worth while to go into details which will help towards our object, which is to hatch a great number of chickens. All eggs placed in the incubator—8,000 or 10,000 monthly in big establishments—not only represent capital because of their intrinsic worth, but because of the potential profit they represent.

Everybody interested in poultry has read and re-read—if not discovered by experiment—that several questions, such as moisture, deaths in shell, crippled chickens, and so on, have not yet been solved satisfactorily ; in fact, artificial incubation hitherto has not succeeded in dethroning the old mother hen. There is with her no question or doubt about moisture

Incubators

ventilation, evaporation of the liquid, etc. She is perfect in these points, especially when not interfered with. Why, then, should not the incubator be as perfect as the hen? This is a question which has occupied my thoughts for a long time, till I made up my mind to study for myself these new and pressing questions. The numerous admonitions, " Do this," " Don't do that," of the expert or amateur were not sufficient to satisfy me, because I could not judge if they were qualified by their experiments to give advice.

Presuming that my English readers are endowed with as much common-sense as one would expect from such practical people, I do not doubt they would like to read some logical explanations of what I have learned. Eventually, they may go further. " Science being other people's knowledge," all of us can add a link to this beautiful chain which makes for progress.

Selecting an Incubator.

For commercial undertaking, the motto " Time is Money " has to be remembered every minute. Where four people have to attend to 30,000 chickens produced annually, saving of time means saving of labour and of money. For this object we must choose an incubator which can be easily and rapidly manipulated. Select one with a 400 egg capacity, fitted with interchangeable drawers, and possessing an easy contrivance for the turning of the eggs.

Our incubator shall be chosen with a nursery, in which newly born chicks will gently crawl towards the glass door attracted by the light, and by their falling into it make room for the other eggs. With the help of this device they never lie on the chipped eggs, nor do they interfere with the coming chicks' efforts to break through the shells. They do not remain in the drawers, and do not, consequently, raise the temperature too much. This automatic removal of the chicks is a great improvement, which saves many lives in big undertakings.

We place our incubators in a well-aired incubator room or cellar, the incubator to be heated and regulated several days to a steady temperature of 100 deg. Fahr. Care should be taken not to raise the heat to 103 deg., for at the beginning of incubation such a temperature is too high. The eggs ought not to be introduced to such a heat, which covers them with moisture. When this dries it causes an evaporation of the liquid contained in the egg, and occasions a desiccation of the shell, which (the shell) undergoes a chemical transformation, subjecting the embryos from the beginning of incubation to an unnatural process.

Imitating Nature.

With the hen there is no rapid starting of incubation when she sits on the cold eggs to brood, her own temperature by the contact of her body with the eggs falls, and it is only by degrees that her eggs absorb the required heat of 103°. Consequently, beneath the hen no sudden change of temperature occurs, and we must follow as closely as possible the living model given to us by Nature. Another example of natural law for a long time puzzled me. How was it that the hen always

Incubators

successfully brought to light chickens from eggs laid some twenty-five or thirty days, while in artificial incubation eggs of ten or fifteen days of age are considered too old to produce vigorous chickens? Long thought and numerous tests have shown me the reason for this strange fact. When the hen lays astray every day or every other day she requires from twenty-five to thirty days to complete a batch of fourteen to fifteen eggs. Nevertheless, all her eggs hatch practically the same day—her chickens are strong and vigorous. The eggs have kept their vitality better than when we preserve them ourselves for artificial incubation. Why? Because the hen sitting on her nest heats them and revitalises the dormant germ by starting in a minute degree the process of incubation—so it is that the older eggs are hatched at the same time as the newer ones, As soon as this fact became clear to me I began experimenting, and found a way to preserve the vitality of the germs of eggs laid in my house in a special apparatus which not only lack of space prevents me describing here, but the necessity of improving it to such simplicity that it might be placed in public hands and worked satisfactorily and surely for commercial purposes. I hope to resume my work in the near future, and complete it and several other appliances.

Having avoided by the above precaution the rapid starting of incubation, we shall see after the introduction of our drawers full of eggs the temperature of our incubator will at first fall. Then, by reason of the partial filling up of the machine and of the decreased cubic space of the air to be heated, as soon as the eggs have absorbed the heat, the temperature will by itself rise to 103deg. If this should not be the case after the lapse of twelve hours, our lamp should be turned up.

Turning the Eggs.

The eggs should be removed from the incubator for turning and cooling every morning and evening, but the same care and precautions should be taken each time not to subject them to a sudden cold atmosphere. It is easy to understand that in taking them out of the machines in which a temperature of 103deg. exists to place them in a room of only 30deg. or 35deg., the sudden cold will provoke too great a contraction of the embryo, which cannot fail to be prejudicial to the vigour of the chickens. To avoid this a piece of canvas should be put over the eggs.

Five or six drawers should be, for the rapidity of manipulation, taken out of the incubator and placed on a long, well-levelled table, and never, as I have so often seen, placed on the top of the incubators. This is neither practical nor convenient, and may subject the machine to movements which are likely to disarrange the regulator. After the removal of four or six drawers, the first one should be uncovered and the eggs turned, and if they are sufficiently cooled put back into the machine. If not, they should be re-covered, this time with a lighter canvas, and the other drawers attended to. Too prolonged a cooling * is also frequently injurious to the embryo.

* This is easily tested by holding the egg against the cheek; there should be no feeling of either cold or warmth.

Incubators

When the broody hen quits her nest no such sudden change occurs. The eggs remain in the deep and well-sheltered nest on the warm bedding of hay or straw. This material retains the heat at the bottom of the nest for some time, and the eggs are gradually cooled.

In artificial incubation cooling is obligatory, were it only to counteract the overheating which might have occurred in the course of incubation, and also to provide the embryo with fresh and pure air.

Ventilation and Moisture.

Airing and moisture are two very important, interesting and very much discussed questions. Although all the incubators are provided with ventilation holes (apertures) to admit the entry and outflow of air, this construction is vicious in most cases. There are a few well-constructed incubators which are less prejudicial than some others, and might be perfect, with small modifications.

To be convinced of this false conception one has only to return to the natural model—the broody hen—and to compare every detail of both ways, artificial and natural, as I did, to find out that there is no reason why the chickens should suffer in their vitality in artificial incubation. To make those comparisons, time, money and patience have been necessary. I hope my readers may benefit from them. Thousands of eggs were incubated by broody hens, and thousands were incubated in several good and reliable incubators—some of them fitted with a tank (hot water), some of them heated by hot air, some having a moisture device, some without—and the results closely compared. All the eggs were of the same origin and freshness, set the same day. They were tested by electricity every day, scores of eggs were sacrificed at every period of their life, and carefully examined, dissected, weighed, etc. In no case were they developing in artificial incubation like those under natural conditions. They were too dry or too full, and the shells subjected to a more severe chemical transformation. The observation of these defects inspired me with the desire to try to remedy them.

Why were they too dry or too full? Must we give them moisture? How is it that with a hen it is not required? Must we admit more or less air into the incubator? Why are incubators so constructed? Is the contaminated air enclosed in an incubator injurious to the chicks? etc.

Results of Experiments.

Here are some answers and results of my experiments and conclusions without going into minute details and too long explanations. No incubator should permit the intake or emission of air. It is the certain cause of the evaporation of the liquid contained in the eggs, as well as a great cause of so many dead in the shells, and crippled chicks. The fact that many constructors introduce a receptacle full of water in the incubator, or other moistening device, shows that they are well aware of this drawback. It is to compensate for the dryness of the atmosphere, subjected to a more or less perpetual draught, that they have to do it. But unless we have recourse to the "air cell test"

Incubators

advocated by American people it is difficult to regulate moisture. When too much is given, the egg, full of liquid, bathes the chick and makes it impossible to move and turn in the shell. If not enough moisture is provided, the air cell gets enlarged, the membrane which is around the chick shrinks and encircles it in such a way that it prevents not only its development but its breaking through the shell. As far as the air cell test is concerned, it requires time for examination and judgment which can be at fault.

What does the hen do in natural incubation? Not only by the natural formation of her body does she, by placing the convexity of her breast into the concavity of the nest, prevent cold air reaching the eggs, but when her nest does not appear to her deep enough, or the eggs on the outside seem too much exposed to the cold, she takes the hay or straw of the nest with her beak, and by a move of her neck from side to side brings her warm bedding over her wings, covering them so completely as to exclude all access of air. Under her there is no draught. Even when she turns her eggs over she does not rise, but rolls herself in the bottom of the nest, thus changing the position of the eggs. We may realise still more that Nature does not desire the evaporation of the liquid in natural incubation, when we see that the eggs coming in contact with the skin of the hen are covered with a thin coating of fat, which renders them bright to the eye and soft to the touch. This glaze fills up the pores of the shell and prevents evaporation when, during the absence of the broody hen, they are exposed to the air. It also prevents the chemical change of the shell so noticeable in artificial incubation, where the eggs are hard and rough to the touch and so brittle do they become that the slightest shock produces cracks which, though often imperceptible to the naked eye, cause the embryos to perish in the shell.

The fact that the constructors find it necessary to build their machines with ventilation holes, although they know the consequences, leads me to think that they believe in the absolute necessity of evacuation of gases and carbonic acid which emanate from the chicks during incubation, and in the necessity of admitting fresh air and of purifying the contaminated atmosphere.

Contaminated Air.

A very striking fact that happened in my house made me determine to thoroughly investigate the question of pure air, for I had the proof, not only by the fact here related, but by numerous experiments, that the question of fresh air had not the importance that is generally thought. Some years ago, before the undertaking of poultry keeping on a paying basis, I had entrusted to my lady companion the care of my last broody hen under which we had just placed a second batch of eggs. Returning from a journey, I went to see the broody hen. As soon as I entered the yard in which she was set, I was struck by a strong and vile odour which infected the whole atmosphere around. An unsound egg must have been broken under the poor creature. She ought to have been moved and the nest cleared out, but the chickens were due that

Incubators

day. I decided to wait, convinced that nothing alive could come out of such an atmosphere, infected, I was told, since the commencement of incubation. Two days later I learned that thirteen fine and vigorous little chickens were walking about with their mother. I had the nest examined, and found beneath the straw a crushed and decomposed chick belonging to the first brood. What, then, of the theory of expulsion of gases, of contaminated air, etc.?

Biology has taught us long ago that the lower the life of the embryo the less is the need of oxygen, but this fact proved it practically. I wanted to have it proved scientifically, and for this purpose I had an analysis made of the air underneath and around the broody hens and, at the same time, of the air enclosed in the incubators, the apertures of some of which for this purpose had been closed, and also the incubator rooms. Much the same gases were discovered, and about the same high percentage of carbonic acid.

If these gases—so poisonous even for human beings—did not hurt the embryo, why drive them out with those injurious and drying draughts? Is it to provide the chicks with air? They must evidently breathe in one way or another, unless admirable Nature has provided for this as well as it has provided for moisture in the eggs. May it not be that either the egg cell contains enough air for the tiny creature, or that the contraction of the chicks exposed to the cold provokes a suction through the porous shell that renews daily its provision of fresh air? Or again that the blood is

Experiments and their Results

óxygenated when cooling by means of the temporary lungs which are placed directly underneath the porous shell. At all events, as the poisonous air is not hurtful, we must conclude that either the chick does not breathe the outer air, or that the porous shell is so constituted as to filter the air of its impurity. I leave it to scientists to demonstrate which of these hypotheses is the most probable, whichever it may be.

Once convinced that the presence of these gases was not injurious, and that underneath the hen the chicks renew or absorb fresh air only during the absence of the mother, I began to experiment with incubators hermetically sealed, allowing the chicks to breathe fresh and pure air only during the time of cooling. The results were such that I continued to close the apertures.

Controversy and Discussions.

After thousands of eggs had been successfully treated in this manner I published the facts. They gave rise to violent and heated discussions in the columns of several papers. My method was tested by many, and by a very old Benedictine, a scientific expert, who used to incubate thousands of eggs annually. He experimented at first with 4,000 eggs, and wrote that he had never before obtained such a high percentage of rapid and easy hatching chickens, without a single faulty bird. The same experiment has just taken place in England with the same beneficial result.

We must bear in mind that in my method, in order to enable the chick to absorb the quantity of

Incubators

fresh air that may be required, my incubator rooms are in a perpetual draught and consequently very well aired. I should never permit such stuffy atmospheres as exist in many incubator rooms.

As soon as the apertures of the incubators are closed draughts are no more to be feared. When in a well-aired room the incubators have fresh air admitted to them, the results cannot be good, as the stream of air dries up the liquid in the egg and alters the shell.

Having also found that in artificial incubation some of the moisture emitted by the eggs is absorbed by the wood of the machine, I usually on the nineteenth day spray water with a special syringe underneath the nurseries, but on no account on to the eggs, in order to compensate for the absorbed moisture. This enables my little birds to be hatched rapidly and easily.

In the natural way the mother hen, when she is permitted to choose a place for a nest, will select a slightly damp spot. At the time of hatching the incubator should not be opened until the twenty-second day. All the chipped eggs remaining should be sacrificed as they are not worth keeping and would never produce a profitable market bird.

Once the hatching is over, every nursery is brought separately to a table, and the tiny birds are placed in a big box, so arranged that 240 chickens go in twelve partitions. Then they are transferred to the brooder house and placed in their rearer in a few seconds. The eggs are tested by electricity on the third and the eleventh days in a dark place situated in the incubation room. Everything is arranged to save time and 400 eggs are tested in eight minutes.

Microbe Dangers.

Two words more before leaving the incubation question. Some incubators, after having been used for some years, are found less regular in their working. It has been attributed either to the regulator or to the dampness having worked into the wood. Certain reasons made me think that was not the case, and after thorough investigation I found out the cause. The wood does, indeed, absorb the moisture. The regulator is not at fault. The cause is much more serious. The dampness goes into the lining stuff of the double wall of the incubators. This warm dampness, combined with the disengagement of gases, encourages the propagation of tiny microbes which work into the wood and fall upon the eggs. These animalculæ make little holes, visible only by means of a very powerful microscope. They work into the shell, producing thousands of infinitesimal holes which destroy quantities of embryos when deep enough. To relate how I found that out would require too much space. The evil discovered, I sought for a remedy, which I obtained by dint of energetic fumigation, killing and eradicating the cause of the evil by this defensive measure. There are, certainly, other ways of removing it. I, for my part, adopted the one described above because it gave me the desired result and at the same time admirably disinfected my incubators.

Incubators

We shall now follow our chickens to brooder house No. 1, and go through their first stage; then study their surroundings and explain some of the causes of English failures to produce table chickens for the market at a profit.

CHAPTER IV

CHICKENS IN THE BROODER HOUSES—I

THE production of table birds at a remunerative profit being our object, we cannot afford to keep our pullets as they are generally kept on the English system. The " hardening off " of chickens, so much advocated by experts and breeders in this country, is very good as far as the production of breeding and laying stock is concerned. We apply it also, but begin it a little later. For the production of table birds, muscle, stamina, vitality and hardness are not sought ; they do not give soft and juicy meat. The hardened birds have not the same propensity to make white, tender and fat flesh, which is what is required, and we shall see later on how contrary to common-sense is the English method, where quality of meat and profit are the principal objects to attain.

Ventilation, not Draught.

The English way of scattering about numbers of little coops or brooders on the wet and cold grass in winter is bad for commercial purposes, as is also the custom of permitting the chicks to sleep in little heated apartments in which there is either insufficient ventilation, or too much. For what is called ventilation is sometimes nothing else but

Chickens in the Brooder Houses

draught opposed to rapid production at a minimum cost. Were it not for the expensive foods that are provided for the chicks by the English method, they would not thrive. In Belgium we do not use food to help the chickens to struggle against the conditions of changeable, wet and cold atmosphere. With us the food must be utilised to make flesh. No ground oats are given to them, unless for breeding purposes. This wastage of food and the tremendous amount of labour required in the English way are part of the cause of failure in this country in the attempt to produce table birds of fine quality commercially and economically. The comparison between the two methods will place my readers in a position to judge for themselves, and to enable them to do so, let us see how my chickens are dealt with after leaving the incubator room.

In big undertakings three brooder houses are generally in use. In the first one the chicks live for the first thirty days of their life; then they go to brooder house No. 2, in which they live a month and pass into the third bigger brooder house, where they remain till they reach eighty or ninety days, at which time they are sold; then the brooder No. 3, being completely cleared out, is thoroughly disinfected, and the chicks from No. 2, hardly noticing the change, take possession of it. The little chicks from No. 1 occupy No. 2, and the newly born chickens are brought from the incubators into the vacant and clean No. 1 brooder; the incubators are left empty for some days and then filled up again with eggs, which are due a month after the preceding hatching. This uninterrupted succession keeps the business going perpetually and smoothly all through the breeding season. There are several other methods of procedure. The chicks may be left in the same big house till they attain market maturity, then the house cleared out and filled up again and new chicks got ready for the following three months; but unless we use three houses the production is checked and the establishment of a regular clientèle prevented.

2,000 Chickens every Three Months.

Whatever the method may be, it is easy to rear on a small scale 2,000 chickens every three months, either by producing them in flocks of seven or eight hundred monthly, brought up in three small brooder houses from which they are not removed, or in big flocks of 2,000 in a bigger house every three months. It makes 8,000 birds annually, which would bring an average profit of about £250. On a big scale the same number of birds reared would bring in at least £320 for every such number of birds.

In large establishments three graduated brooder houses are needed. It is much better, on account of the regularity of the output, to produce monthly and to have the birds changed from house to house every month. I have seen brooder houses in England in which birds of different ages are kept, but they are not reared on the same principle; we shall see how complicated such management would be on a big scale and how far from being as healthy as our method. The best way to realise these differences and to have a more striking and illuminative view of the method of scientific establishments is to follow our 240 chickens when they leave the incubator

Chickens in the Brooder Houses

room and are brought to their first brooder house, where they are placed very quickly in a big rearer. The roomy box from which they have been transferred has had its twelve partitions previously lined with square pieces of soft woolly blanket. By taking these square pieces of blanket out by the four corners they are gently placed in the rearer without any of the chickens being killed or crushed in the transfer. They find there ample accommodation and plenty of top air without the slightest draught. This rectangular rearer is warmed by means of a lamp * placed in the centre, whose reflected heat on the tin top of the cover provides a sufficient temperature. The lamp is surrounded by a screen of perforated metal covered with a black material, no light being permitted to show in the night and tire the little tender eyes; for a perpetual light is not only injurious to the eyes, but to the nerves, and consequently to the whole constitution. Nature has made darkness for rest. Under the hens there is no lamp to cause the chicks suffering, and in the daytime they find the restful effect of darkness underneath their mother. To imitate this beneficent obscurity the glass tops of their rearers in my establishment are darkened several times in the day by dark green blinds; the glass top is not covered at night in order to permit the last inspection of the day, which takes place at 9 p.m. A glance into the rearer shows how the chickens behave. They are generally collected a little way from the lamp.

* Fitted with double flat wicks. An ordinary powerful lamp will do, if it allows for the extinction of one wick as the chicks grow older.

Training the Chicks

Should they be too far away the heat is too great and the lamp is turned down; if near the lamp the temperature is too low and the heat should be increased. The chicks are trained from the very first day of their life to come out of their rearers when they are called, by means of a tapping on the wood, and to go in by a clapping of the hands. It is really marvellous and charming to see, after two days, how obedient the little birds are. The rearer is placed in the sleeping compartment, its floor raised about four inches above the ground. Three little front doors permit the exit of the chicks; these doors are closed the first two nights and afterwards replaced by ones of wire netting. All along the front length of the rearer there is a slightly sloping board, with some strips of wood nailed on it, making a very convenient stairway. Access to this rearer is so easy that from the very first day the chicks come in and out of it quite by themselves, without the risk of their being unable to find their way back; it has no corner where chicks could crowd together, no underneath part where they could hide themselves. The rearer is about 8 ft. long by 6 ft. wide. Each rearer is separated from the other by wire netting.

The first two days their litter is made of hay, afterwards to be replaced by finely ground peat moss and covered with a thick washable, soft cotton cloth, stretched and fixed at the corners. This cloth is changed every morning and night when the lamps are attended to, and thoroughly washed and disinfected. A small litter of chaff mixed with sand and cut clover and bran is spread over it, in which they enjoy scratching.

Chickens in the Brooder Houses

Ventilation Shafts.

The big brooder house has two compartments—the sleeping quarters and the recreation room. In some of the English and American brooder houses there is also a division generally made of wire netting; but this does not fulfil the same purpose at all, as we shall see. In my establishment, which is the most modern of those existing in Belgium, the compartments are separated by a brick wall, not to make a corridor in which the rearers are placed, but in order to give them two distinct places in which to live. My birds do not sleep in the same atmosphere or temperature in which they have been during the day. Their sleeping compartment is not heated, it is aired by a constant admission of outer air passing through a double wall. The air is attracted by means of ventilating shafts placed at the top of the wall separating the sleeping compartment from the recreation room. The ventilation is regulated in cold weather by closing one or two of the numerous shafts; the air of the houses by means of this feature is constantly renewed.

Strengthening the Lungs.

Birds should not sleep in a warm room as they do in so many Belgian and American establishments. Where a great number of chickens sleep together they raise the temperature of the rearer sufficiently and it would be dangerous if the sleeping compartment were heated. When birds have been too long a time in a high temperature they are found panting in the morning and do not recover from this over-heating. I want them also to purify and strengthen their lungs; and one of my ways of keeping them healthy is by giving them plenty of pure oxygen, much space, light and scratching exercise. This cannot be attained when they remain day and night in the same house on the same litter, breathing the air contaminated by their droppings and by the lamps. Two compartments provide a change of surroundings. They run from one place to another, seem to enjoy life better, and are kept more active. In winter, the nights being long, they pass half of their lives in their rearers and in the daytime they frequently return to them to warm themselves; as we shall see later on, every time they go to the sleeping compartment they find themselves in a cold and pure atmosphere. As soon as they are three or four days old the running space allowed them is increased. Access to the recreation room is permitted; there the windows are continually open.

Birds may be kept in large numbers in the same inside brooder without the necessity, as so many people imagine, for close stuffy houses; no hothouse for them in my establishment. The brooder house is large—30 ft. wide (internal measurement), 15 ft. high, and in two lengths of 80 ft., divided by a store and administrative room 14 ft. by 39 ft.

The sleeping compartment is 12 ft. deep and is divided into spaces 8 ft. wide. The rearers placed in it should be about 7 ft. or 8 ft. by 6 ft. This gives the chicks for the first few days a space of about 6 ft. deep outside the rearer in which to run about, till they are permitted to reach the recreation room; this last one measures 18 ft. deep and is divided by

Chickens in the Brooder Houses

easily movable and adjustable wire partitions into spaces from 5 ft. to 8 ft. wide. The placing of these partitions depends on varying circumstances. The recreation room is heated by means of four big pipes running over the high doors leading from the recreation room to the sleeping compartment. The wall opposite the one that cuts the brooder house in two parts is made partly of glass; the return pipes run underneath this glass front.

Concrete and Asphalt Flooring.

The heating is of the same sort as generally used for horticultural purposes, and may be regulated at will. The temperature is kept at 16deg. to 18deg. Centigrade (i.e., about 61deg. to 65deg. Fahrenheit), which is about the same heat the chicks would experience were they hatched in spring. Not only are all the windows left open, but in the centre of the roof there is a series of ventilating chimneys that may be opened at will. Attached to this big brooder are outside yards in which the birds are permitted to run whenever the weather permits; in these runs fir trees and shrubs give sufficient protection and shade. The floors of the brooder house are made of concrete, the walls are coated with cement plastering, the sleeping compartment has an asphalt floor. The ground slopes very slightly towards gratings tightly covered by sheets of iron to prevent air and smells coming into the house; six or eight such openings are placed the length of the house.

The washing and disinfecting water is collected by an outside drain which carries it far away from the houses. The twenty runs of the brooder are provided with a water tap, the water falling into a long gutter, which allows a large number of birds to drink easily together on both sides. The outlet of water from this gutter is by means of a pipe which joins the drain ; a plug keeps the water in during the day. These gutters are movable and new ones set in place every day, the dirty ones as well as the dirty trough being taken away every evening and soaked in water during the night in a long concrete trough into which disinfectant has been added. In the morning they are brushed and placed on shelves. It takes more time to explain all these precautionary measures than to do them, and it has not been possible to show all these little details of equipment on the plan, which is intended only to indicate the general arrangements of the building. The roof is covered with Eternit and plastered underneath. All windows are made to lift up inside the building and near the roof—thus preventing the birds perching on them, and stretched wire netting prevents them flying outside—the glass top roof does not open.

Early Days.

The birds in such well-lighted and well-kept brooder houses are very vigorous and cheerful. It is a charming sight to see them rushing out of their rearers at the sound of my voice, chirping so happily and contentedly, that they are the admiration of the very few who are permitted to visit them. For the first eight days they seldom leave their sleeping quarters. After their meals have been given to them, they are allowed one hour to play, then they

Chickens in the Brooder Houses

are ordered to go to bed by the clapping of hands. The doors of the rearer are closed, the blinds pulled down and, after some minutes, the most complete silence reigns—they sleep, digest and rest. An hour and a half afterwards the blinds are pulled up, the doors opened; they come out and immediately rush to the mash which has been in the meantime prepared for them and placed in troughs. It is very striking that though after eight days they are no longer ordered to rest, yet by force of training they have got so much into the habit of doing so that they keep fairly to their old custom later on in life. This promotes quick growing and makes them beautifully fit for the fattening pens. Chickens newly born are like young babies; eating and sleeping are most necessary for them, too much playing about exhausts and excites them very much; they do not realise the want of rest; it is important to regulate them in this way. Do not tell me that it is too much fussing about; it is certainly less than the moving of coops and the running about from one to another. I found it was well in my own interest to care for and study their well-being thoroughly. Sentiment allies itself in this case with profit.

Comparisons with English Systems.

As such a brooder house contains from 4,000 to 6,000 chickens they have to be taken care of every month and are worth the trouble devoted to them. Such a business is, in fact, a manufactory of chickens. How could these operations be carried out in the English method? Where fifty or sixty birds are kept in each separate brooder it would be quite impossible; it would require more time for thirty brooders to be attended to than for the thousands of birds that are kept in our houses. Our chickens are much better cared for, better cleaned, better looked after than by the English system; they are much happier and healthier than when kept in the cold wet and wind of the open air of December or February. They are too young to know what suits them during the first days of their lives; but by our method they never catch cold, because they are not exposed to damp or wind, also they have no vermin. We shall see later on how important it is to keep them free from these pests. I know how strange it must appear to the English breeder that we keep our birds together in such great numbers; we can assure him that we preserve many little lives by our method which would have been lost on the English plan. Like English breeders we have some mortality, which occurs generally in the first days of their lives, at the time when the chicks have cost very little; but the advantage of cleanliness, of rest, and the good effect of absorbing pure air in the long nights and the artificial long day that we provide for them, as well as the saving of labour, and the effectiveness of severe supervision, give us such a great margin of profit that we should eventually be able to afford the losses caused by crowding. Those losses occur sometimes in some badly managed houses that we know of where 450 birds are kept together, but in our own establishment we did not find the percentage of mortality important at all; it is certainly less than among those experimented with by other methods.

Chickens in the Brooder Houses

The saving of many lives and the quickest growth are two important considerations to take into account. The more birds that a brooder house contains the more profit it brings, and the quicker the birds are ready for sale the better it is for our purse. By my method of feeding and keeping them I very often obtain a difference of eight days or more in their growth. The multiplication of those days gained on thousands of birds economises thousands of days' food and brings in very high and unexpected profits. Added to those preceding advantages only one lamp is needed for each 240 chickens, that is one fourth the quantity of petroleum that would be necessary where they are kept in flocks of fifty. Remember that only one rearer has to be cleaned instead of five, that one litter is sufficient, etc., and you will realise the economical worth of my methods, which I claim to be more healthy, practical and scientific than those applied up to now in the intensive culture of table-birds.

When birds are reared in little coops on the grass in little sleeping apartments not thoroughly cleaned every day, where constant supervision would be difficult if not impossible, they cannot always escape roup, gapes, diphtheria, etc., all consequences of cold, damp, wind and dirt ; they can never be absolutely free from vermin. The people of our country know what it means for their pockets to put birds free from these pests in the fattening coops, and are most particular about it.

Chickens of different Ages.

I would like English people to see the difference

between a bird reared in January, February, or March by the English method, compared with the same bird of the same age, brought up in my way, with no vermin and no checking in growth. They would find my bird twice as big as theirs. Another cause of failure in the attempt to rear table-birds on a big scale profitably and commercially in England has been in the keeping together of chickens of different ages in the same building; it is neither practical logical, nor healthy. The odour which emanates from tiny chicks, even if they are very well aired and cleanly kept is as much as their lungs can bear; but if they have to breathe the emanation of bigger ones night and day, it is too much for them. It should only be allowed where birds are brought up in little flocks of fifty or seventy-five, in a room which allows a vigorous inflow of air perpetually where they are given free access all day long to open runs. I have seen several poultry farms in England where birds are kept in this way at the beginning of spring for breeding and laying purposes, but it does not answer so well where birds are reared for the table.

How is it possible to obtain a uniformity of attendance, a business carried on automatically as in our way, where everything is simplified to such an extent that once the business is started it goes by itself? If birds of different ages are kept in the same house, what sort of control shall we need in order that such things as cleanliness, heat and food are ensured as they ought to be? The mixing together of birds of different ages is thoroughly bad. The temperature which is suitable the first day of life for a newly born chick is far too high for a bird of twelve or

Chickens in the Brooder Houses

fifteen days. The food—mostly wet mash for table-birds—which suits the first age does not suit the second one, and if different sorts of food have to be ready, according to the different ages of birds, they cannot be made so economically as when we have only one food for 1,000 birds. The distribution of these different foods would also complicate the attendance and supervision.

The hardy way of bringing up the chicks which is in vogue in this country is very good for breeding purposes, but would be still better if it began only when the chicks have passed through their first tender days. So many little chicks are handicapped from the beginning which should be able to pick up strength and keep alive, if properly treated at starting. I was told so often that they were not worth keeping. It is childish. Keep them if you can, were it only for the table. I am afraid it will be a difficult business to change English minds on this subject, but I will try hard. People here are not easy to move. I found it out myself in many conversations that I have had on poultry matters, but if I succeed in inducing them to try, I know that they will soon agree with me. Unfortunately, if there are people eager to be taught, there are still more that do not even desire to consider or try other methods. I cannot quite explain this, except by a want of deduction, an obstinate " sticking-up " for the old ways, that is absolutely contrary to intelligence and science. If all that which has always been done has to continue for ever, I wonder where and when will improvement and progress be made.

CHAPTER V

CHICKENS IN THE BROODER HOUSES—II

THE three brooder houses which are necessary to make an establishment for the consecutive rearing of table-birds are increased in size according to the age of the birds. When the chicks are a month old they are transferred to brooder house No. 2, * where they find more room and larger rearers. These are kept cold for the night after a fortnight, since the birds at six weeks generate enough heat when in large numbers. The rearers continue to be heated in the daytime, in case some of the birds may still require warmth, but after two other weeks they do not need any more artificial heat, except in the recreation rooms which are heated continuously in the day time.

The birds remain in No. 2 brooder house for thirty days, that is, until they go to brooder house No. 3 ; this last one is still heated, but in a less degree. In this house the birds do not have a rearer to sleep in, but roost on a slatted board through which the dirt passes, and no more litter is given to them. Every morning these boards are scraped, and the floor thoroughly cleansed. The night manure is never

* This house measures 150 feet, against 120 feet for brooder house No. 1, and 180 feet for No. 3.

Chickens in the Brooder Houses

permitted to remain in the house. The boards are ranged along the walls, thus freeing the full space of the sleeping quarters. The birds are thus hardened to stand the cold, to which they are not at all sensitive.

It is generally considered in poultry keeping that winter chickens do not develop as quickly as in the spring. I could not detect the slightest difference with my method. The brooder houses in my establishment are fitted with electric light. If birds reared for table had to go through the long winter nights without eating, they would naturally not develop and mature as quickly and easily as in the spring, when nature gives them long days and short nights. When dusk falls, birds go to sleep; tiny chicks having tiny stomachs need to be fed often; were they left till 7 a.m. without food, they would not develop so consistently. For this reason I had to lengthen the day with the help of electricity, and I found it paid to do so. Even birds of six or seven weeks old would not attain full table growth rapidly had they to fast for such a long period. In the afternoon, at 5.30, the dark houses are lighted. The birds think it is daytime again and come out of their rearers to find their meal ready. They eat, drink, etc., and three-quarters of an hour after the light is turned off in the recreation room in which they are usually busy scratching after their meal. Suddenly they find themselves in the dark, but fortunately for their distress, the sleeping compartment in its turn is lighted up, and through the openings, which are provided in the communicating walls for their exit, they see the light. What a relief! They rush through the openings and find themselves again in their sleeping compartment ; but no joy being everlasting, the lights are gently lowered. How quickly night comes ! it is quite time to go to bed again. The little and the middle-sized birds return to their rearers, the larger ones take their place again on the board, and gradually the light fails, till complete darkness reigns. At 8 p.m. the same process begins for the last meal of the day, and an hour after all lights are extinguished, this time till the following day. They have already had two extra meals. In the morning, at 5 o'clock, the houses are again lighted. A new day has begun. They come out to receive their first meal, then they are gradually put in the dark again, and they return to sleep, this time till daylight, which comes about 7.30 a.m. When the birds come out of their rearers for their meal they find their recreation room sweet and fresh, the windows having been wide open all night long. The heating apparatus has been attended to, for this is the first work of the day. The windows have been closed for a while to permit the room to regain its regular heat, then opened again for the whole day.

The Importance of Light.

From the store and machinery rooms the electricity can be switched on to light the brooder houses for the first and two last meals of the chickens, without having to go there expressly, and from my private house I can also direct all the lights. Plenty of light being necessary for health, and being the greatest enemy of microbes, the front of the houses

Chickens in the Brooder Houses

and part of the roof are made of glass. The light is increased by the whitewashed walls, so that the brooder houses are as light as the open. At first the birds were very happy in them, but later in the season the sun began to shine; then I experienced some of the difficulties that people who keep birds in greenhouses discover, without being able to detect the reason of this check. Birds appearing perfectly healthy died suddenly, until at last I realised that the sun was the cause of the trouble, and remembered not only that too much strong light was an excitant of the organism, but also that the sun passing through glass has an increased burning effect. Paper can be burnt if placed behind a piece of glass exposed to the rays of the sun. The chickens were simply dying from congestion of the brain. I had to diminish the light of the house, and thought to do so by means of curtains, but they are not only a harbourage for dirt and smell, but soon look shabby, so I use them only when I cannot manage otherwise. After different experiments I found the right way to eradicate the trouble. I put blue into the whitewash that was used for the walls and painted the windows with it. Immediately the houses took an appearance of peacefulness and freshness. I knew the soothing effect of blue, and that it would intercept the exciting rays of the sun. After having applied this measure my birds were perfectly happy and healthy. Of course, I was aware of the disastrous effect of the sun on young chicks, and took great care not to allow them big runs when they were newly born, in order not only that they should not have too much space in which to exhaust themselves, but that they should not expose themselves to be burnt by the sun, of which they are very fond. Those deaths occurring in the brooder houses No. 2 and No. 3, in which the middle and bigger birds were kept, immediately aroused my attention. At this age, unless through accident, I experienced very few deaths.

The Workers and their Treatment.

To be successful, an establishment of such importance has to be managed with mathematical precision, and everybody working in it has to be interested in the success of the business. All the people I employ receive a percentage of profits on the condition that the mortality does not exceed a certain amount. The larger the losses in the house the smaller the percentage they receive. Consequently it is necessary for one man to have the whole responsibility and the management of the house under his care. From the attendant's room (storeroom in the plans) the whole length of the house can be seen through windows provided for this purpose. All doors are double-action swing doors, which help the attendants greatly. The coal-cellar is just behind the attendant's room; thus the man has the heating apparatus always at hand. The washing and disinfecting of troughs, gutters and cloths are done in a big concrete trough, which is placed there for that purpose. Nothing else, except an iron barrel of petrol, some baskets for the carriage of chickens, shelves on the wall, a zinc table for the filling of the lamps, different utensils, and sacks of peat-moss litter necessary for the daily renewal of

Chickens in the Brooder Houses

the litter in the rearer, is in this room; no food is anywhere near. The man receives from the machinery room, and at the right time, only what has to be distributed in troughs.

As an important feature of control a book is hung on the wall, in which all mortality is to be entered. All dead birds have to be shown, their left feet cut off, the feet placed on the table, counted, then taken away. This prevents the man presenting the same bird again on another occasion. The deaths of big birds are generally to be attributed to carelessness. Birds are caught in the doors, some have their legs broken, because they have been hurt by heavy buckets, stepped on, etc. These birds are good for the table, but the smaller birds who die through other causes have to be burnt immediately.

The sale of mature birds has to correspond to the number entered in the house, mortality being subtracted. It is easy to realise that without the strictest control thefts can be committed, and I found the above prevention a very good one.

To go into the numerous and very important details absolutely necessary in such a business would lead me too far. It should always be remembered that thoroughness of detail, constant supervision, a sense of organisation and the strictest economy are obligatory if large profits are to be obtained.

Unsatisfactory Foster-mothers.

Nothing being perfect in this world, this business as well as many others has its drawbacks. Among the material difficulties of such an undertaking there is one which is very important, namely, the choice of a rearer. Many foster-mothers (rearers) are on the English market, some of them very good; but not one, as far as I know, is really practical and economical for commercial purposes. They are made for too small a number of birds.

Some establishments in Belgium, England and America have tried the system of hot-water pipes running some inches above the floor and covered with little coops, around which small curtains hang; but this was not and could not be a success. Underneath these little rearer boxes the poor little chicks have insufficient ventilation and very often too much heat; only the lucky birds who are near the curtains can breathe freely; the others are very often suffocated. With this system they are kept in the same house; they sleep on the ground, where they may be subjected to draughts, and they breathe the whole day and night the vicious atmosphere of the room. They have their little beaks close to the litter, in which dirt accumulates every day. There can be no question of changing the whole litter of the house, for it would be too expensive. (By the way, litter in such an undertaking is always a very expensive item.) By this system the heat cannot be properly regulated. In short, nothing in it is wholesome or practical, and nowadays it has been discarded nearly everywhere.

New American coal-heated rearers are now making their appearance in England, which are very much spoken of in America. I have seen some of them lately. They are all of them made on the same principle, to rear some hundreds or thousands of chicks,

Chickens in the Brooder Houses

but they have also many drawbacks. In fact, I very much doubt if the ideal rearer for an establishment on a big scale exists yet, either in my establishment or in any other. I possess, as far as I know, one of the best for commercial undertakings. In 1914 I was busy with the construction of a big rearer having the practical requirements, *i.e.*, economy, facility of attention, hygienic principles, no central heating, etc., like the small ones I had improved and utilised for nine years; but the war broke out, and my plans to construct big commercial rearers had to be abandoned instantaneously.

The Temperature and Management of Foster-mothers.

In some establishments in Belgium, birds are kept in their rearers as well as in the brooder houses in a terrific temperature. It makes them so unhappy that the poor creatures, in order to get relief and permit the heat to escape from their bodies, have to raise their feathers. They are kept so hot that they are too exhausted to move. These birds are hothouse birds. Birds should never be overheated in their rearers and houses; it not only makes them mopey and unhealthy, but they risk contracting illness when exposed to the slightest cold. Chickens coming from such stuffy establishments cannot even bear the transfer to the fattening pens, where they arrive with colds in the head and purple skin. The heat must be calculated to give them the same temperature that they would experience in the spring.

In some other establishments I have seen chickens sleeping in square closed boxes which I regard as

56

suffocating coffins. Whenever a rise in temperature has taken place during the night, pails of dead chicks are taken away in the morning from these boxes. How the remaining chickens manage to escape death is a problem to me.

In another Belgian establishment the complicated exit of the chicks from the rearer is not to be relied on, and is so impracticable that a lot of little ones can never find their way in again, and exhaust themselves in climbing up the long, steep stairs. In spite of these deficiencies the business in these establishments still pays, which evidently shows that our methods, even if not perfectly applied, have money in them.

Difficulties to be Overcome.

I have heard of many failures which would not have arisen if people were careful enough not to throw their money into such a business, utterly ignorant of the difficulties of the work. Apart from the material difficulties which have to be thoroughly mastered, moral assets are necessary such as perseverance, will, energy, patience, early rising, hard work. There must be no week-ends, no Sundays, no holidays, and, to close the enumeration of what I consider the most important factor of success, there must be love of the birds, without which everything would be too much trouble and too hard. A day of neglect may be a set back in the health and in the growth of the chicks. The incubators, rearers, lamps, as well as the distribution of food, and the heating apparatus, have to be attended to every day; supervision can never be relaxed; attendants

Chickens in the Brooder Houses

can never be trusted; there is too big a stake to risk through the carelessness of anyone. We know already that when two or three thousand birds are in every house it means eight to ten thousand in the three brooder houses, which represent a large amount of money. Many people imagine that three months' tuition on a poultry farm is enough to enable them to undertake poultry on a paying basis, and to master the multifarious details involved in it. No greater error can be made. I am not astonished to hear that such people have not succeeded; I should indeed be much more astonished to hear that they have done so.

The rearing and keeping of a bird, of a living creature, all the year round, in good and continuous remunerative condition, is surely as difficult as to make, let us say, a gentleman's hat, or a pair of boots. For these things an apprenticeship of some years is accepted as necessary. The same apprenticeship should not be too long in a school, where the individual tuition amounts to some minutes daily, to thoroughly learn a business which is so highly profitable in skilful hands. I wish I could impress my readers better than by logical deductions or by my personal experiences. So long as no modern establishment is started in England, so long as English people will not see the need for change, I am afraid it will not be very easy to induce them to try other methods. Although the best friend is the one that is the most sincere, I know that truth and criticism are sometimes unpleasant to hear or to read, and national habits are not easily changed. I wish I myself had received such advice and criticism when I began poultry keeping. How many expensive stupidities I should have avoided; how many more years of experience I should have had. Unfortunately, I did not meet a sincere friend who would dare to tell me that I was wrong. I had to find it out myself, and I know at what cost. That is why I dwell upon possible errors that can be made and have to be avoided in starting poultry rearing.

Belgian and English Methods.

Might I be permitted to say that no establishment in Belgium can bear comparison with mine? As for the Flemish peasant, he rears his chickens in winter in the kitchen. He has no place in scientific poultry keeping. The other poultry establishments are conducted on the old principles. The birds sleep in the same room, on the same litter, in a continuously warm atmosphere, etc. Many of these establishments have been constructed little by little and, on that account, cannot be worked scientifically. It was only after having visited the older ones and seen myself the improvements I should have to make that I realised all the requirements of such a business. It was necessary to learn how to save labour, economise in work, prevent contagious illness, promote health, and make the business more profitable. Then I built my establishment and worked it absolutely scientifically, which permits me to-day to call it, with reason, "The rearing of chickens scientifically improved." I do not mean to say that everything in it is perfect; far from it, it could be better. Experience has already taught me many things that I

Chickens in the Brooder Houses

should, in order to make still larger profits, change, add, or suppress, were I to start anew.

In our big rearing establishments the fowls are sold by the hundred couples together for cash. The rearing of chickens on a big scale can only be done by people of a certain class on account of the investment of money necessary. The difference in customs between England and Belgium forbids our gentlefolk fattening their birds. In their establishments it would not be approved of. For this reason the clientèle of big establishments consists especially of fatteners, who come to buy the birds, put them in fattening coops, kill, truss, and send them to their own customers. It is not difficult to find a market for the birds. As soon as the fatteners know where to find fowls free from vermin, they are only too pleased to come and fetch them, even if they have to travel for them. Before the installation of such rearing undertakings they had and still have, as in England, higglers, who go about and collect for them, from the peasants, the birds of the special breed " Grey Coucou de Malines," which is the only breed asked for in the market in Belgium. Naturally the birds collected in this way have vermin and sometimes roup or other ailments. Those that are ill are generally killed without being placed in fattening coops and sold to the happily ignorant buyer. The others are fattened. As for ridding the birds of vermin, there is no hope. Therefore it is not possible to expect a bird that is eaten up by parasites, increasing every day, and from which it cannot rid itself, to utilise the food provided for it to the best advantage of the fatteners.

Precautions against Diseases

Rest and other Factors in Fattening.

Rest is one of the greatest factors in the development of fat. These clever people soon realised the profit that clean birds would bring them. They became very eager to buy where they knew the birds were kept in good condition. As we do not undertake delivery, they generally come with their own baskets. The chickens to be sold are shown to them, but before entering the rooms the buyers have to put on shoes belonging to the establishment, because they may come from places which are not free from vermin and diseases, and every precaution has to be taken against the spreading of these. All visitors have to conform to the same rules. The birds are picked up, placed in our own baskets, which are carried to the threshold, where they are taken away, counted again, and placed in the buyers' baskets. Strange baskets are never permitted to be carried to the house, as they also may bring vermin or illness.

Few of my birds, I have been told, have to be placed in fattening coops; they are generally found in such good condition of flesh and fat that hardly any of them have to undergo the fattening process. They are killed directly and sold. This particular advantage, which brought me so many orders that I could not expect to be able to execute them, is attributed to my way of keeping and feeding birds.

In order to work satisfactorily either for the fatteners or for the market, a breed, "a unique breed," has to be selected to make good table fowls. It is one of the first things that English people will

Chickens in the Brooder Houses

have to realise. So long as they will not abandon the thousand extraordinary crosses they breed, and so long as they continue to rear birds for table in such irrational ways, they will never succeed in producing a good quality of table fowl with profit.

CHAPTER VI

THE FAULTS IN ENGLISH TABLE BIRDS

I HAVE tried already to convince some breeders I visited that they were spoiling the quality of their table fowls by their crossings with birds of Asiatic origin, and that they were putting on the market by this procedure a number of inferior birds. I urged them to take in hand again, and breed exclusively from, their good and beautiful Sussex, Kent or Surrey fowls, but to my numerous objections I was told that the crosses made with Indian game birds increased size and gave more breast meat and hardiness. I know that, but nothing can be so detrimental to table production as a cross with such birds. I tried to show them that the size obtained from this cross was synonymous with coarseness of skin, quality and colour; that the breast meat of such a cross was of a dry, short, fibrous texture; that the legs were too tough because of too much muscle; that the skin of the legs was dark in colour; and that the produce of such crossings lost the juicy, tender and white flesh of good birds. Weight has never been a test of excellence, and a medium bird, every morsel of it good, would be a less expensive one in the end, even if more had to be paid for it. The legs, I was told, were given to the kitchen. In

63

The Faults in English Table Birds

France the legs are very much appreciated; many people prefer them to the breast, as being less dry, but juicy and tender. I told them that French and Belgian people would never accept those high, coarse, yellow birds; that here, as well as in France, the really good birds were the white-legged and white-skinned ones; that the Surrey fowl and the beautiful Dorking (this last one always regretted by the poulterers, who complained to me of its disappearance) were the birds to breed from for table purposes; that to do so would be one of the ways to become self-sufficing, and avoid importing the first quality of table fowl from France and Belgium. I told them that thousands more of good quality birds were always wanted here; that there were never enough. I told them that there were already too many of that low class bird on the market, on account of the egg production obtained in this country, especially from birds of yellow legs. I added that the standard, the quality of table fowl, was falling every day; that it was really useless to lower it by breeding for the table from these birds.

The Crossing Mania.

I have just returned from Sussex, where I have been, at the invitation of the Board of Agriculture, to visit several fattening establishments. I saw there, among the beautiful Sussex, numbers of inferior Irish birds. Evidently, if there were enough good fowls, the fattener would not take the trouble to fatten these birds. They cost him the same food and labour as good ones, and certainly do not bring in the same profit. Supposing that for hardiness

64

crossing is absolutely necessary, good white-legged birds could be crossed together. I told them that the old English White-legged Game would answer this purpose splendidly; that the French faveroles —imported direct from France, without having passed through the hands of the fancier, who improved it by increasing the size, but spoiled the quality—would make a splendid cross, etc.

Among the many replies that were made during those conversations, I heard several times: "But we are advised every day by poultry experts to cross with these birds." Alas! it is but too true, and it is a great pity. I quoted to them a saying of Mr. Harrison Weir, a famous and clever English breeder. "No greater curse," said he, "could have come to our homestead poultry than the persisting, crazy howl for cross breeding."

I was also told that American people have no objection to yellow-skinned birds, but, on the contrary, preferred them. It may be so, but American people are far from being good judges in the way of cooking. Apart from rich people who can afford to pay high wages to their cooks, their meals are mostly composed of *delikatessen* and preserves, and for my part I prefer to rely on French taste. Such crosses do not produce the uniformity of production, the evenness in growth and in quality that is required in a commercial undertaking.

I know people are very often prejudiced against the opinion of a foreigner. I wish my readers to realise there is no *parti-pris*, no peculiar mood of mine in my criticism. I want them to do better than is done now, to understand the logic of my

The Faults in English Table Birds

explanations. I know the business of which I am speaking thoroughly, and I am able to tell them the mistakes to be avoided, the errors that they make when, for example, they couple together the production of eggs from yellow-legged birds with production for the table from the same birds.

Mr. Harrison Weir's View.

To give more weight to my assertions, here again are the opinions of Mr. Harrison Weir, whose book has been deeply interesting to me. In his accurately informed work, *Our Poultry*, he himself quotes many opinions of celebrated poulterers and breeders which strengthen and confirm my own. It is highly regrettable that his serious advice, written fifteen years ago, has not been taken more into consideration. "At Leadenhall Market now," said Mr. Harrison Weir (speaking of the quality of the actual bird on the market compared with previous years), "the change is but too apparent, and the diversity of form, colour, coarseness of flesh and fat, with shanks and skin of varied tints from light to dark, the former often black, brown, green, blue and yellow, and each with a different anatomy from the next; but when the non-apreciating public cry for size and bones, the higglers and fatters will, of course, so trade as to meet such demands." Further, speaking of the nearly disappeared excellent Dorking table fowl, he writes: "One of the greatest evils that befell the large, well-formed, active table fowl of the Southern Counties was the introduction of the Shanghais or Cochins. The higglers brought to our homesteads cockerels of the new breeds in exchange for the old time-honoured birds, and prevailed on many of the Kent and Sussex farmers to cross them with the great old fowls that before were the perfection of the barndoor breeds. The beautiful Dorking and the English Game fowl are the true British poultry. They are racy of the soil, and come down to us, like many other good things, from a remote antiquity. If it were possible to engraft the hardihood and quality of the latter upon the size and early maturity of the former, perfection would be obtained. The veriest gourmand would ask no more, for there would be quality and quantity to satisfy the most capacious and capricious of appetites. Tenderness and plumpness would go hand in hand with juiciness. Fifty years ago the Dorking was of such superb excellence that it had achieved for itself a world-wide reputation. This is now contradicted by the modern breeder; but, if it was not so, why was it eulogised by everyone, and pronounced by the unbiassed thinking men as perfect? No other fowl of its size could compete with it for sterling merit as a table fowl, and it was by no means despicable as an egg producer. . . . One of the best sitters and mothers, its chickens coming early to maturity, and even while growing, fattening easily."

Later on, writing about the Kent, Sussex and Surrey fowls, he said: " Why cross them? Why not have used every precaution to keep the blood pure and true? A bird that, with its congener, the old English game cock, has been the pride and boast of our English farmsteads for hundreds and hundreds of years. Years and years ago I protested

The Faults in English Table Birds

against the reckless way in which foreign crosses were introduced, and not only that, but by the advice of persons said to be authorities in respect of pure and proper methods of breeding poultry and other kinds of living stock ; and also that the prizes were still awarded to breed unmistakably cross-bred, these being bought by the unwary and ignorant, as far as poultry was concerned, as pure bred, who were told and urged to cross again in another direction. So confusion has grown more confounded."

Mr. Harrison Weir also writes in his book : " Edward Bond, the greatest naturalist, when he saw the first Cochin, said : ' There is the ruin of our English breeds of fowls, for credulity always lacks sense ' ; and never were his words more true or exemplified than now."

Keep the Breeds Pure.

How much I wish my English readers would meditate over these sensible, and, alas ! too true quotations. It is time, indeed, to take serious measures. From these extracts we see that the evil does not date from to-day. Of course, we all know that the pure breed of the past and present day have been the result of several, but, happily, better crossings made in remote times ; but once a breed has been fixed, do not introduce any more combinations of strange blood, which will too often bring unknown and detrimental elements and crowd the market with bad table birds. If only those crosses were made by breeders possessing knowledge of natural laws, and not by the blundering, unscientific modern and hasty method of procedure in the way of crossing and recrossing with bigger, coarser and dryer-fleshed birds, the peril would not be so great ; but, unfortunately, it is not the case. The result is a terrible mixture of birds, whose several crossings can no longer be traced by their owner.

I am afraid this mania of making new breeds or new crosses is already too deeply rooted in the English brain to be easily changed. A bird should be chosen for table production, of any kind preferred so long as it possesses the quality of a table fowl ; but once chosen, stick to it. This bird should not have been in the hands of the fancier, and no exhibition points should intervene. It should be bred regardless of the colour or the artificial points of the standard, because all birds have been, and are still being, more or less spoilt for practical purposes by observing these points. Such crossings should be avoided in commercial undertakings, not only because they complicate the work, but because the introduction of an inferior fowl, for the sake of increasing breast-meat or size, can only spoil the quality. *The introduction of an inferior bird will never make a better one.* This is so logical that one would think it required no explanation.

The wrong way to produce *Petits poussins*

To satisfy the German clientèle, who wanted a bird of large size, we had, in Belgium, to increase the size of our grey Coucou de Malines. We did not introduce an inferior bird for this, but crossed with the Combatant de Bruges, or Bruges Game, which possesses white skin and legs, and is very similar in

The Faults in English Table Birds

quality to the old English Game. Belgian people, in order to make a big bird, do not have recourse to a yellow-legged one. The lack of comprehension of what a table bird should be is to me incredible. I saw several times how the *petits poussins* were produced in this country.

In Hampshire, where I went to visit some poultry farms, I came across a gentleman, who was intelligent and very keen on poultry, devoting his time to, and obtaining his living from, the production of *petits poussins*. For this purpose he was breeding from white Leghorns. His little birds were very well kept, and charming to look at. But the breed is quite unsuitable, and of inferior quality for the production of milk chicken, asparagus chicken or Hamburg chicken—all those small birds which are called in France *petits poussins*. How can it be possible for people to realise that they are doing wrongly when they read, as I did myself, advice in poultry papers every day to breed *petits poussins* from the Leghorn breed. Such advice should not be given to ever-credulous, because very often ignorant, readers. This class of production (*petits poussins*) requires better elements in order to be profitable, and I am sorry to say it is no more understood by many people than the production of a good average table fowl. We get it quite differently, and it is a very remunerative product. We produce it from the cross of a Breakel cockerel or Campine, with our Coucou de Malines hens. The same cockerels crossed with the Sussex fowl, or, better still, if possible to find them, with Dorking hens, would breed a superior and early maturing chicken.

Food Wastage

Supposing for one moment the Dorkings of to-day lack vitality (which I do not believe, otherwise the breed would not have held its own for centuries), the introduction of Breakel blood imported direct from Belgium would give it to them. The desired results can be obtained with any other table fowl possessing white legs and of early maturity, but our crossing makes a quickly feathered, plump and superior little bird, sold generally at four or five weeks old, according to the weight required. Thousands of these chickens are in demand in Belgium and in France; a single house in Brussels asked me for 40,000 of them yearly. I had to decline, because I keep to the chickens in which I specialise.

We feed our *petits poussins* and our chickens for table in quite a different way to those bred for laying or breeding purposes. In England people do not make any difference in the feeding. We are not so wasteful, and we know how to bring them to maturity in a much quicker way with a far better result. The Leghorn *poussins*, to which I referred above, were just half the size they would have been by our method of feeding.

The First Feeding.

My chickens coming out of the incubator room are brought to the brooder house No. 1, where they remain thirty-six hours without receiving food. Were they born, for example, on Monday night they would receive nothing till the Wednesday morning. Close study concerning the time they have to fast has proved to me that this number of hours was necessary to permit them to digest the yolk of the egg

The Faults in English Table Birds

that Nature has provided for them in the shell. Their first meal is given to them on the white wooden stairs of the rearer. Those interested in poultry matters know that the chick in the shell develops from the white part of the egg, but an ordinary person does not know this. The mistake of many articles is that they are generally written for those who know, and consequently are very often incomprehensible to those who are only learners. The yolk of the egg remains intact in the shell till the moment of hatching. It has to be the first food of the little bird. For this reason it is always a dangerous process to help the chick to come out of the shell. If the absorption of the yolk which takes place through the umbilical cord is not completely done, the chick will die from this intervention; and if the little creature receives food before the complete digestion of the yolk, it begins life already handicapped. Thousands of chickens are killed every year through fear of their fasting too long. The ideal and most complete food for them is the yolk, which contains protein, fat, lecithin, etc.

When I began poultry keeping I knew no better than to copy everything, good or bad, that was printed; but, helped by my great love for the birds, I soon realised that the food which I gave them was not in many cases what it ought to be. Some chickens develop bowel troubles, enteritis, white diarrhœa, etc. The food I used was a sort of custard made with infertile eggs, to which milk, bread, greenstuffs and sand were added. It appeared to me very strange to have to give them this sort of food, no hen ever having cooked eggs for her brood. But it was printed, and I thought I could do no better than follow the directions, though several times I asked myself whether I should not try something else. My doubts were suddenly confirmed by a very interesting study and illustrations of a dissected chicken that I found in the American *Reliable Poultry Journal*, in which the deaths of the chickens were attributed to the agglutinated cooked albuminoid of the egg, *i.e.*, the white part of the egg. Particles of this matter were found undigested in the bowels, and they were the cause of the inflammation. In their tender days all chicks have not the same dissolving power of the gastric juice. Those that were lacking in this respect became ill. It was enough to enlighten me. From that very day I resolved to study what would be the most suitable food for them and never trust books or advice any more. I plunged into chemical feeding, and I found there invaluable treasures, some of which I am pleased to share with my readers. The peasants as well as the commercial establishments for the rearing of chickens in Belgium, give them the custard referred to above, and continue to do so. I could not make up my mind to suppress the giving of eggs completely, because of the valuable effect of the lecithin on the feathering, and the fat and protein matter of the yolk; I knew that Nature wanted them to have it. So I separated the yolk from the white, and as the white, being an albuminous matter, was too valuable a food to be lost, I gave it with great benefit to the young pigs.

The Faults in English Table Birds

A Feeding Experiment and its Results.

I made different tests, four of which are given below. These tests were made on hundreds of birds, and every test for several months, which means that thousands of birds have been subjected to them. The chickens were kept under the same conditions, the same number were placed in a similarly constructed rearer; the temperature, the litter, the exposure, the age, the quality and the breed of the birds being absolutely the same, and all of them in the same brooder house.

Pen No. 1 received the ordinary custard (with the exception of the white of the egg, which I had omitted) cooked with milk, to which brown bread, greenstuff and sand were added. Pen No. 2 received brown bread soaked in milk, the superfluous milk being afterwards squeezed out with the hands, greenstuff and sand. Pen No. 3 received brown bread only, finely granulated, no milk, also with the addition of greenstuff and sand. Pen No. 4 received the yolk of the egg, hard boiled and then passed through a sieve in order to make tiny particles of the yolk, greenstuff, sand and no milk. Then I watched the results.

As soon as my little birds saw the food on the stairs—as they were very hungry they had splendid appetites—they began to eat. I could not detect at first the slightest difference. The second day in my establishment the food was always given in troughs; then the results began to appear. The birds of pen No. 4 were most particular to pick out the little yellow particles, and only when there were no more did they eat the bread and greenstuff. This indicated to me their greediness for it. The result was that sometimes the bread remained in the troughs. It will be noticed that this food is nearly the same as pen No. 1, with the exception of milk in the food. Milk is given to them to drink. It was absolutely wonderful to see how quickly they developed and feathered. They were eating more of the egg than the first ones, these being obliged to eat bread with it. The food left over in the troughs always being removed, it is evident that the food left over in pen No. 1 had in it a certain proportion of egg that had not been eaten. On the contrary, in pen No. 4 what was taken away was bread and no egg. As soon as I realised their fondness for the yolk, I found out what I wanted to know, namely, that they could eat a great quantity of yolk, without having the slightest bowel trouble.

I knew that I could continue with benefit to give egg without fear, which is not the case in the other way, when the white part of the egg is cooked in the custard. The birds were in splendid health. For ten days I continued to feed them in this way ; then, for the sake of economy and other reasons which will appear later on, I replaced the bread with a part of what is, I think, called sharps or pollard. An equal quantity of barley meal was added to replace the bread, which was little by little discontinued, as well as the egg. The barley meal was sifted, the husks being removed for five days. The birds at the end of this time being a fortnight old, the barley meal was no longer sifted. In our country it is not so finely ground as in England.

CHAPTER VII

TO ENLARGE THE YOUNG CROP

An important object in feeding young chicks is to encourage the enlargement of the crop. As soon as they are a fortnight old we do not object, as in England, to particles of unground husk in the mash; it gives bulk to it and slowly distends the crop, permitting the birds to absorb a greater quantity of food, which consequently produces a quicker growth and a greater propensity to get fat.

Uses of Water and Milk.

The experiments made in pens No. 2 and No. 3 showed that it was better to moisten the bread with milk than to give the milk as a drink. Milk standing too long is liable to turn sour and to scour the birds. Bowel troubles are always to be avoided in tiny chicks. Milk is a food by itself, and its casein matter is immediately coagulated in the stomach by a ferment (rennet) which is present in the gastric juice. Another ferment (lipase) acts upon the fats contained in the milk, splitting them up into free fatty acids and glycerine. The work of gastric juices to which milk is subjected would be an additional strain on the bird's stomach, if it were given as a drink the whole day. On the other hand

Studying Various Methods

it does not quench their thirst as pure water would. Water not only washes the kidneys, but is also a conveyer of the nutrients in the blood and lymphatic vessels. Milk should be used only in the mash, this one being made and eaten in a few minutes; it will increase the protein ratio and economise part of this costly matter. When milk is given in the food the lactic acid quickens the digesting of it, but, on the contrary, when absorbed in constant and great quantities, it disturbs the digesting organs and prevents the food from being properly utilised.

Pollard and Barley Meal.

As soon as I had selected from these and many other experiments the food I intended to give to my birds, I started my tour of rearing establishments, with the view of studying the different methods used to produce table birds. I had adopted pollard and barley meal because I knew these meals to be the general basal ration of commercial undertakings. I wanted to see the effect of this food myself, leaving the study of it to the future, already having resolved to change it if I found it unsuitable. Wherever I went nothing was used but these meals, the pollard being increased slowly to two parts as against one of barley meal. Meat meal and phosphate were added to the mash in the proportion of 5 per cent. of each and increased till it reached 8 per cent. when the birds were eight weeks old. The mash in some establishments is mixed with separated milk, in some others simply with water. It has a soft consistency, not crumbly as it is given to hens in England, nor so watery as it is given in the fatten-

To Enlarge the Young Crop

ing establishments. Pure water is given to drink, and fresh green food is placed in nets high enough to oblige the birds to jump to reach it. When there is a shortage of fresh green food, cabbages or beet-roots are hung in the pens instead. In some establishments when the birds are ten weeks old a bucket of buckwheat meal takes the place of one bucket of pollard. Nothing else is given till maturity, with the exception of tiny cracked grains spread in the litter to induce them to scratch and develop their muscles as well as their appetites. This grain, in the proportion of about 4 lb. to 100 birds, is given to them from one to four weeks old. The mash is distributed in troughs placed over the litter and fresh mash is never added to the old. This is the very simple Flemish way of rearing table birds for commercial purposes. The birds are fed every two hours till maturity. Pollard is certainly not only a cheaper meal than ground oats and chicken food, but as the proportion of barley meal is small and fattens the birds more quickly, making them beautifully fit for the table, it makes a very economical mixture and pays well.

Feed Young Things Little and Often.

In my establishment also the birds are fed every two hours from wooden troughs, and receive just what they can finish at once. From six to nine weeks old they are fed every two and a half hours, and from nine weeks till their departure from the house every three and a half hours. Science has proved many a time that the food is assimilated better when little is given at a time; little and often as long as they are building up their frames, and less when they are older, when it is better to lengthen the time between the meals as the birds do not require so much food for growing, having already attained the required size. The troughs are placed on the asphalt floor of the cold sleeping apartment, where the birds are obliged to go for their meals. There is no litter there, as I found it most unappetising to feed them over litter more or less dirty, and the asphalt floor is swept every day.

Economy in Feeding.

I soon realised, thanks to this cleanliness, the appreciable economy I should be able to make in the distribution of food. Birds are very wasteful creatures and they throw a great quantity of the food given to them into the litter. On my smooth asphalt floor I saw at once the quantity of food trampled upon, and as I wanted to stop this wastefulness, constructed a special guard for the troughs after having tried several other devices.

I covered the V-shaped ordinary trough with a higher and larger inverted V-shaped slatted board, the pointed part of this large V coming over the sides of the 3 in. deep tray, which is placed underneath the trough, and over it the food falls when the birds pick it out with their beaks. The slatted wood bars are meant to prevent them throwing the food out by a side movement of their heads. On the top of the inverted trough a thick loose wire-netting is spread, preventing the birds from perching on it, and soiling the food or the trough. (By the way, birds are never allowed to perch when they are

To Enlarge the Young Crop

reared for the table.)   After the distribution of food, the empty troughs are collected ; then, the birds having no more to eat, are very pleased to pick up from the tray all the fallen bits. This apparently insignificant and simple detail has saved an average of 50 lbs. of dry food daily. I also found that it was better for the sake of economy to send from the store room in little buckets the exact quantity of mash to be given—so many troughs to fill, so many little buckets are sent full of mash. They have to be sent back empty immediately to the store room to prove that they have been distributed at the time. It was absolutely impossible to get the attendants to give as little as I wanted, as they were convinced that the more they gave the birds the quicker they would develop. This proved untrue scientifically. It is always better to divide the daily ration—that is, the quantity that is supposed to be necessary for the day—into numerous meals, for the heat which is produced during and after feeding is then divided more regularly. The productive value of the aliments is also greater when the food is spread over a certain period of time. This has been proved in fattening experiments when the same quantity of food was given at one time and when it was divided into several meals. Grit is not usually given in rearing establishments, because it has been noticed that birds fed on soft mash eat very little of it. Grit is useful for the crushing of hard grains as well as for supplying the birds with mineral substances. This last point is generally overlooked. The omission of grit, unfortunately, is not made up as it ought to be. The birds are deprived of the mineral substances so useful to them and which are generally dissolved and completely assimilated by their organism. This lack of mineral salts—potash, soda, sulphur, lime, magnesia, oxide of iron, calcium, etc.—which are contained in greater or less quantities in grit, is the cause of the trouble from which they suffer, as we shall see later. There is no doubt that wet mash is the best of food for the production of table birds, for whenever their growth has been compared with that of birds fed on grain, a striking difference has shown itself.

Weak Legs and Phosphate of Lime.

But there is a drawback to this rapid growth, for the heavy weight which the chickens attain while they are so young is too much for their tender bones. Their legs cannot carry their heavy bodies. Some of the birds are so weak on the legs that they cannot keep standing, while others are unable to walk to their feeding troughs, and so get no food. Birds so affected are generally the best developed of the lot. After some days of partial fasting they naturally deteriorate. To prevent this, as soon as this weakness appears they are killed, though sometimes they are too small to bring in the same profit as they would if they had attained maturity. If they are not killed, this state of weakness will increase daily in spite of the greatest proportion of phosphate then given. It will be useful for my readers to know that unless they get the real precipitate phosphate of lime, which is very expensive and not always reliable, many of the phosphates sold for feeding purposes are nothing but valueless materials, such

To Enlarge the Young Crop

as bone meal and bone ashes. The assimilation of these two last phosphates is no more than 13 per cent. to 14 per cent. It is even very doubtful if we obtain from the precipitate phosphates of lime sold to us the amount of advantage we are entitled to expect, for it is a well-known fact that, four or five times more than is necessary has to be given in order that the bird should retain a part of it in its body. Let us say also that should real phosphate of lime be given it might sometimes be found dangerous to use it, as it may contain quantities of arsenic and sulphuric acid, because the latter is now used in some places instead of hydrochloric acid for dissolving out the mineral matter from the bones, which is used to make the real phosphate of lime. It will be still more useful to know that the greater part of the phosphates sold for feeding purposes come from mineral sources, in which case the phosphate is not assimilated at all. It is money thrown away and a useless strain on the birds' organs. The best phosphates are those extracted from vegetables. The German chemists have put these very valuable dry phosphates on the market, for human use, under several names. Their extraction is a costly process, therefore prohibitive for animals. Realising that the troubles of the birds came from lack of phosphates or salts I had no rest till I found assimilable phosphates and mineral salts suitable for eradicating these bone weaknesses that had also appeared in my house, as well as another disadvantage which is very much complained about in the rearing establishments. I am pleased to say that I succeeded beyond all my expectations in curing both. (The recipe for this phosphate preparation will be given in a later article.) This second drawback is a cessation of growth that attacks the birds when they are three or four weeks old and sometimes reappears between the ages of eight and nine weeks. Curiously enough these drawbacks, weakness of the legs, and cessation in growth, exist also—not to such an extent, naturally—in England, where I did not expect to find these deficiencies in birds fed mostly on cereals. I know that it is so, as different people have complained of it to me—among them a successful breeder of great repute.

The Need of Scientific Feeding.

I had been greatly struck at seeing in France, Holland, Belgium and Germany that the producers of table fowls, who also experienced these troubles, did not even attempt to change their methods. This *laissez faire* indolence was unbearable to me, and though at the time my scientific knowledge was more theoretical than practical, I undertook to change this bad state of things for the better. I knew enough of the laws of digestion to realise that the uniform and scientifically incomplete feeding given for three months was most harmful. I knew the properties and the component parts of food, and could compose a ratio. That was about all my knowledge. I had read many interesting facts, as well as the results of experiments carried out in France, and especially in Germany, but had been rather discouraged when I found that these experiments were made for the purpose of promoting

To Enlarge the Young Crop

the higher production and well-being of cattle, sheep, pigs and horses, but that very few things had been tried for poor, neglected poultry. Fortuately, I knew that certain fixed laws could be applied to different animals; then with this little, but already valuable knowledge, I started to work for my birds, trusting to the practical and thorough experiments I intended to make (to which I would add logical deductions and observations) to compensate for lack of real scientific knowledge. I taxed my memory for all that I had stored up, and applied myself again to scientific and chemical feeding. I do not intend to enter into more scientific details than are necessary to elucidate what I have to show. If I say to my readers that a food has a ratio of 1 : 4 or 1 : 10 I shall have told them nothing, for what does it really mean to say that a ratio of meal or fodder is made of one part of protein to four or ten parts of fat and hydrocarbonaceous food? Does it give the productivity of this food, its digestibility and usefully assimilative proportion? Very few people, indeed, are able to make a mixture of food knowing exactly the proportion of protein, crude fibre, etc., contained in it. All that has very little real practical value, because the secret of feeding lies in the assimilation of food. For example, a food might be taken in and expelled from the body having been only partially or perhaps not at all assimilated. It does not enable us to detect the exact productive value of it. In the making of a ratio the injudicious introduction of a food, or too much of it, to balance it may lower its digestibility, consequently altering the most intelligent calculations and leading into error. Practical experiments alone determine the most suitable and the most paying aliments, for every sort of animal does not extract from the same food the same amount of nourishment.

The Process of Digestion.

Although I am eager to avoid too long and dry explanations, it will be found useful to go a little into the process of digestion for those who wish to understand better the requirements of their birds. It will show them that feeding is not so simple as it looks, and that the stomach and the different gastric juices play an important *rôle* in the digestion of food. Under the term " digestion " are included all those processes by which the substances contained in the food are converted into a form suitable for assimilation or absorption. Food is transformed by five different juices which act upon it during its passage through the alimentary canal—(1) saliva, (2) gastric juice, (3) bile, (4) pancreatic juice and (5) intestinal juice. The chief work performed by the saliva consists in the softening of the dried foodstuff before reaching the stomach, where the gastric juice, which contains free hydrochloric acid as well as lactic acid, acts upon the proteins and fats of the nutrients. The food is now partly digested. It passes then from the stomach into the small intestine, where it is mixed with two other digestive fluids—the bile and the pancreatic juice. The bile plays an important part in the digestion of the fats, for it dissolves a large quantity of the fatty acids coming undigested from the stomach, as well as emulsifying,

To Enlarge the Young Crop

that is, dividing into minute particles, the unchanged fat of the food. It also stimulates the muscles of the small intestine and increases its movements. The pancreatic juice in its turn exerts a powerful digestive action upon the proteins as well as upon the fats and starch. The partially digested food next meets the intestinal juice, the effect of which is to change any remaining protein or starch in a manner similar to that exercised by the pancreatic juice. To the action of these various digestive fluids must be added that of the bacteria which play an important part in the large intestine.

All the productive assimilable substances extracted from the aliments are absorbed by capillary vessels which carry them into the blood.

Variety in Food.

This brief and incomplete outline of the changes which the food undergoes in digestion shows that the process is not limited to one organ. For this reason the more varied the food, the nearer it will approach the ideal, because foods are digested principally in the stomach, others in the small intestine, while others again undergo the chief digestion in the large one. Therefore a mixture of several foods will spread the work of digestion better over the different parts of the digestive track and, for this reason, a varied ration is more suitable than one made from a large quantity of a few materials. The palatableness of a food must also be studied, specially in fattening. To improve it a little quantity of salt should be added to the mash, no more than we should use for ourselves. Salt increases the flow of digestive juices, it promotes activity of the circulation and prevents disturbances of the digestive apparatus. Too great a quantity would have a contrary effect; we must remember this and also that whenever a change of food is made it should always be gradual; even a sudden increase in the volume of the food may cause disturbance. On the contrary, when a gradual increase is practised the digestive organs slowly adapt themselves and expand by means of the growth of their walls. Knowing these things, I could not follow blindly the methods of the commercial rearing establishments. To alter them meant extensive experiments, which could not be improvised, and would not show their results at the exact moment I wanted them. Results only appear in time; so I had at first to put off this complicated study of a better composition of food and simply try to improve the Flemish ways.

CHAPTER VIII

THE SEARCH FOR A SCIENTIFIC FOOD

**Experiments difficult while the Business
is being Established.**

AT the beginning of my undertaking I fed my birds
on the commercial ration, to which, to increase its
value and its palatableness, were added several foods
I had tried years ago. They were chosen so as not
to raise the cost of the birds, as the equivalent of
food was taken away from the original ration.
These additions were, of course, intended to improve
the quality of table birds. It is not practicable to
make complicated experiments when a business has
to pay immediately, on account of the general expenses running every day. Expense and difficulties
must be faced. There is, at first, an enormous
wastage, and many costly false steps, as well as an
expensive *mise au point* which requires all one's
attention if a profit from the first year's undertaking
is desired. Therefore it is essential to adopt at
once a practical and irrevocable rule, this being the
only way to succeed. It permits one's thoughts to
be concentrated on the same object, and to perceive
what is wrong or could be better.

As in a paying undertaking one has to be most
careful, I waited till the business was in good going

order before beginning to experiment anew. Then I started several tests, carried out for months, with five pens, each containing 720 birds, separated into three lots. The plan of these experiments was to feed a constant basal ration, with the addition of different feeding materials possessing great digestibility and productibility as well as very little crude fibre. One of them was made with a scientifically composed food for chickens. I did not at first believe very much in such foods, because many of them, generally sold under names attractive but not accurately descriptive of the material contained in them, are often made of nothing else than a mixture of by-products of little value. Also, though these foods are sometimes sold with a guarantee as to the amount of protein, it is far from being sufficient, as many materials rich in protein, are not always made of *digestible* protein. Consequently this guarantee is not worth much. To inspire confidence certain condiments, with a smell of drugs, are added to such foods, avowedly to increase their digestibility. Scientific and impartial experiments have shown that these condiments neither improve the effect nor the value of the food.

A Trial of Spratt's.

But once I was determined to see for myself if really something in the way of a scientific food existed for chickens I was most careful to get it from a reliable house. I applied to the English house, Spratt, renowned for its products. One of my pens was fed with nothing but this food. One lot received it in a dry form, the second one mixed with

The Search for a Scientific Food

skimmed milk, and the third with water, to which two tablespoonfuls of linseed meal to the litre were added; the other pens were treated in the same way. Although sceptical as to the result I could not help noticing that, after a time, the birds under the compound food experiment increased and feathered very quickly, and were better and bigger, especially those which had linseed water to drink, than the birds in some of the other pens. The chickens at the time were delicate. An illness of the ovarian organ of the hen (owing to the stupidities of peasants in-breeding too much, etc.) had spread over the whole of Belgium, and though in my house all precautions are taken against illness, I also had more mortality than usual. The poor little birds were already ill in the egg and those hatching had in them the germs of the illness from which later on they were to suffer.

The Spratt chickens showed a lower percentage of mortality, so I had to conclude that the food was more suitable to the birds' organisms, probably because of its variety of substances. During the course of these early experiments I continued to feed them with it, notwithstanding the price, which, at first, I thought prohibitive for a commercial undertaking. Later on I came to the conclusion—when I realised that it had added only 10 centimes (not quite a penny) to each bird's cost for the three weeks that this food was given—that such a small increase was of very little importance, especially if it had saved life. The death of 300 or 400 birds monthly is not in itself a great loss in a commercial undertaking, on account of the very little value of the birds during their first month. The worst is, that their deaths do not decrease the running expenses of the business in the least degree; on the contrary, it raises the price of the living ones. Nothing, therefore, ought to be spared to help them in this delicate period of their lives. It matters little if they cost more, so long as they grow up and bring us a profit, instead of a loss, and a consequent rise in the price of the living ones. After I had made these experiments I was enlightened as to the necessary addition to the basal ration. The composition food tests, having been a practical illustration of the advantages of various ingredients in the food, I then began to experiment in another direction in order to find a scientific concentrated food for the production of table fowls, which would eradicate leg weakness and the stoppage in growth.

The Discovery of a New Food.

After two years of numerous tests, sacrificing and analysing hundreds of birds, I finally found a food which began to prepare the birds for table from the beginning of their lives. In the rearing of any animal, the direction that the animal will take later has to be considered; as those which have to be fattened have to be helped to attain quicker development (in my case propensity to put on flesh easily) by a liberal and suitable diet from the start. The more quickly an animal grows the more concentrated the food should be, and everyone knows how quickly poultry mature compared to the length of their lives.

So my efforts were directed towards a food

adapted to the birds' crops, containing a very great amount of highly digestible nutrients, and calculated to form the maximum of tissue. This food has been condensed in such a way that the small quantity they can at first absorb will contain sufficient aliments. It is very rich in assimilable phosphoric acid, lime, potash, soda, oxide of iron, magnesia, manganese, chlorine, etc., extracted from cereals. As it is a well-known fact that animals do not retain in the body all the mineral and vegetable salts given to them, so the food has to be studied to give the birds a quantity which will not tire their organism, and at the same time will be sufficient. Potash, soda, lime and magnesia are fixed in about the same quantities in the body, whereas phosphoric acid is held back in increasing quantity, which means that the more the animal grows the more it retains of it. Once completely developed it does not retain it any more. A scientific ration is a very complicated business, as it must be studied to fulfil the requirements of the growing animal. The filling up of the crop has to be considered, besides the necessary amount of nutrients. The food must also contain a proportion of digestible crude fibre, not only to give bulk and enlarge the crop, but to facilitate the mechanical work of the intestines. I wanted this food, too, to counteract the tendency that some birds have to anæmia and tuberculosis; therefore, by a judicious introduction of assimilable iron I enriched the hæmoglobin in the blood. An animal which has enough of it will never be subject to tuberculosis. It is anæmia which opens the way to bacilli, by the deficiency of hæmoglobin in the blood. Such a great improvement in the development of cattle has been brought about by concentrated food, and it has proved so profitable for infants, that there is no reason why a concentrated food should not be good for baby chicks. It goes without saying that the preparation of this food can only be done by people fully acquainted with the value of food stuffs and their manipulation. In all probability my formula will be manufactured in England within a short time.

Trials of a New Food.

As many of my readers may have, as I myself have had, a prejudice against compound foods, the profitable adjuncts to the commercial ration on which I have fed my birds previously will be found in this article. They were made to increase the value and productiveness of the uniform and unscientific Flemish ration. In the feeding of fowls, either for commercial purposes or in other poultry businesses, the protein ration generally given is too high. My birds, as the result of experiments, receive a ration higher in fat and hydro-carbonaceous food, for I found that young birds require more fat, as well as a greater quantity of bone-forming food. They need also plenty of heat-producing material, obtained only by fat and carbonaceous foods, as they make proportionately more bone and fat than flesh. It is really wonderful how little heat, if kept in this way the small chicks require. What they need is the internal warmth generated in their own bodies. This is a great factor of their well-being, as with it they can bear a colder and more changeable temperature with greater facility. All healthy young

The Search for a Scientific Food

ones are very fat, and science has proved lately that they can absorb and digest a greater quantity of fat than was thought possible formerly. If the cost of digestible protein were about the same as digestible carbo-hydrates (hydro-carbonaceous food) it would be of very little importance if more protein were given than necessary. It must be remembered that a ration high in protein food, if used in excess, is liable to overtax the kidneys, and even to intoxicate the animal—as protein without exception performs the function of carbo-hydrates and fat. But protein matter is considerably more expensive than fat and carbonaceous food, so we have to use it with economy. A glance into the marvellous science of feeding will show us that the birds find in their ration, even if not rich in protein matters, a sufficient quantity of it. The birds are able to extract protein from the non-protein substances they receive with the help of bacteria, which change them into protein in the intestines. The bacteria increase the decomposition of protein matter, enabling, consequently, a greater proportion of it to be free for the making of body protein. This fact being clear, I strongly advise my readers to reduce the proportio of the meat meal given in the standard commercial ration to a proportion of 2 per cent. instead of the 5 per cent. advocated in Flemish feeding. With the skimmed milk, which is used in the mash during the first six weeks, eggs and other materials, the birds will have sufficient protein. Afterwards, when they are about five weeks old, the meat meal should be gradually increased until it attains a proportion of 5 per cent. when the birds are eight weeks old. At this time a ration higher in protein becomes necessary to develop the fibre of the flesh and enable them to store up a larger quantity of fat. This increase of protein reduces the stoppage in growth that always occurs at this period of their lives. In the meantime it stimulates their organisms in the most satisfactory manner. Later, when the birds reach nine and a half weeks old, it will be well to decrease the meat meal gradually, in order that they receive only a proportion of 2 per cent. till their departure from the house. Not only does the decreasing of meat meal make the ratio higher in hydro-carbonaceous food, but the introduction of buckwheat meal instead of pollard induces them to put on fat splendidly.

The Flemish ration can be improved by the addition of the following foodstuffs given alternately. The five rations will serve as models. One foodstuff may be changed for another without altering its value.

Two spoonfuls of common cod-liver oil are introduced to these additions for every 100 birds.

1. Five per cent, of cooked cracked rice, which is cheaper than whole rice, with 5 per cent. of cooked crushed wheat.

2. Five per cent. of cooked *peeled* potatoes and 5 per cent. of cooked crushed wheat.

3. Five per cent. of cooked maize meal and 5 per cent. of cooked crushed wheat, maize to be given only until the birds are five weeks old, as it would alter the quality of the flesh.

4. Five per cent. of cooked crushed buckwheat with 5 per cent. of cooked potatoes, or 5 per cent.

of rice meal with five *entire* eggs for 100 birds, to be given when the birds are between six and seven weeks old; at this age they are able to digest the albumen of the egg. The eggs are to be either hard boiled and passed through a sieve, or given raw and well beaten, and then added to the rice or potatoes. It will be found that the lecithin ($C_{42}H_{84}NO_9P$) contained in the eggs helps them greatly towards feathering. In all these varied foods a minced sugar beet cooked with the food will be relished by, and beneficial to, the birds. A little salt should not be forgotten.

Such changes excite and stimulate their appetites and are no extra cost to the commercial feeding, be-because 10 per cent. of these stuffs take the place of 10 per cent. subtracted from the basal ration. The mash should be mixed three times a day for various reasons, but specially on account of the skimmed milk contained in it, in which germs of disease may find an excellent *terrain de culture* for the development of microbes brought in by dust, etc. As microbes rapidly increase in virulence, the food has always to be made freshly, otherwise it may cause sickness and even death. When the birds are six weeks old the milk must be slowly decreased in such a way that what is saved in the sixth week may be spread over the first days of the seventh week.

If milk were not such a watery product (75 per cent. at least is water) it would be highly beneficial for partially making up for the lack of vegetable and mineral salts in the mash. Unfortunately iron is very scantily represented in milk, and this product is so diluted that small birds cannot extract enough

Mineral and Vegetable Salts

of the phosphoric acid, lime, etc., that it contains.
The birds require assimilable phosphates, vegetable
and mineral salts. Their organisms appear to
literally crave for them. These mineral and vegetable salts are lacking in all animals. Calves, even
if fed exclusively on the milk meant for them
by nature, show an eagerness for these substances
which demonstrates undoubtedly that they do not
receive the quantity required by their organisms.
They always try to eat chalk, mortar, or other
mineral substances.

This deficiency appears to be general, for we see
it also in human beings. It indicates that our foods
and the foods given to animals, are lacking in these
precious substances, probably on account of the
deficiencies of the soil. In the natural rearing of
chickens, when they wander about with their mother
they add to their diet much material which contains
these important salts. Whether the chicks are
reared for breeding or laying purposes, or for table,
it is most essential to help them to feather and
develop rapidly, especially if they are reared for
breeding, as then they need a bigger frame, greater
stamina and richness of blood. We shall obtain all
that with quickness of production by giving them a
large quantity of assimilable phosphates, mineral
and vegetable salts extracted by a distillation of
cereals—the recipe of it will be found in *Egg Production*. It has a wonderful effect on the vigour and
laying power of the hens as well as on the moult.
In the Flemish method a small proportion of grain
is spread on the litter to encourage the birds to
scratch, as it is only when the limbs are used that

The Search for a Scientific Food

the muscles and bones mature satisfactorily. I constructed for this purpose several covered boxes about 3 feet by 2 feet and 8 inches deep, raised 6 inches from the floor, placed in the recreation room. These boxes are partially filled with chaff, sand, cut straw, etc., on which the grain is spread. Twice in the morning and twice in the afternoon the boxes are opened, the little birds jump into them and scratch to their hearts' content with wonderful vigour. The cover of the boxes is V-shaped, with a loose wire spread on the top of the V similar to the guard troughs. This jumping into the grain, running into the sleeping compartment to eat, and the running to the recreation room again to drink, give them sufficient exercise.

A Little Secret.

I must give to my readers a little secret which discovers the reason of the splendid assimilation and fine flesh of my fowls. Among the numerous books I have read on scientific feeding, I came across some very interesting tests made on geese by a German experimental station. The most important article of diet used in these experiments was charcoal; not given in little quantities to birds as in England, but given in enormous quantity to the geese under experiment. The animals subjected to the test increased in weight, quality and fat in a wonderful manner. Naturally I began to study this food in connection with chickens, giving it to them in a granulated form and in an automatic hopper, from which they could eat it *ad libitum*. The result was that they never had enough of it, and the quantity

98

required to satisfy them was so great that I could not keep up with the crushing of charcoal. Having realised already the economy and the effect on the birds, I bought a special crusher for nitrates, worked with an engine, as are all my other machines. This crusher separated the granules from the powder. A certain quantity of the powder was added to the mash—with a view to better disguising the composition of my food—and was soon increased to two buckets against one of meal. Charcoal is known in the human pharmacopœia as a splendid filter and disinfectant of the bowels. It prevents intestinal disorders and the formation of gases. So I had no fear of hurting the birds in giving such a quantity; I trusted to their instinct and was right. They never were in better health nor of better quality. Charcoal is nothing more than burned wood. Lately the Germans have found the feeding value of the pulp of trees, but the nutrients of the wood being transformed by carbonisation appear to be of very little value as a food. The beneficial effect of it must be attributed to a better extraction or fixation of the nutrients. Whatever it may be, the improvement of the birds was so marvellous that I am convinced I owe to charcoal the fact that the greater part of my birds could be killed and sold without having to pass through the fattening pens. Accordingly I was paid a better price for them than that quoted in the balance sheet, which represents the average price of the birds bought in similar establishments.

The following balance sheet requires some explanation. The feeding of the birds has been

The Search for a Scientific Food

11 kilos. of dry matter each. The average weight of a newly born chick is about 40 grammes. It weighs when sold at twelve weeks old (sometimes, as in my case, at eleven weeks) a minimum average weight of 1 kilo. 750 grammes. All the foods—meal, green, charcoal, phosphates—have been reckoned at an average price of 180 fr. the 1,000 kilos. Although the greater part of these ingredients does not cost this price—the meat meal and oil cost more—some will make up for the others. An average of half a pint of milk has been given daily to each bird for six weeks, which amounts to 22½ pints per bird. As far as I can make out, food and wages are about the same cost in both countries.

The wages of a manager, and the percentage allowed to him, amount to £240 yearly, which could be avoided when the business is supervised by the proprietor. I find that a responsible head man costs me half the price of a manager. If a manager is obligatory, I would, for my part, choose someone with a money interest in the success of the business. I consider this most essential. I would also like my readers to realise how profitable is the breeding of pigs as a consumer of waste products. From the distillation of cereals we get an enormous quantity of valuable food for them. The addled eggs, the dead-in-shell chickens, the 6,000 dead chickens, the refuse of the mash, to which the grass of the land must be added, all make an excellent diet for pigs. The profit shown in the balance sheet is much below the expectation of what can be obtained. To be on the right side, I put the profit as low as possible, and reckoned full expenses. The number of birds is produced from October till the end of May or June.

In such an undertaking in England other trades should be added to the production of table fowls—as soon as the laying birds should be selected, many more eggs should be obtained from them, which will bring more chickens; these may be sold as day-old chicks—or the eggs sold for sitting—in both cases an extra profit would be obtained.

PRICE OF THE CHICKENS AND EXPENSES OF A MODERN REARING ESTABLISHMENT.

EXPENSES.

	Francs.	s.	d.
72,000 eggs bought yearly, of which 50 per cent. chickens are hatched = 36,000 (6,000 of these die). One chicken costs two eggs at average price of 0.20fr. = 1·92d. . .	0.40	0	3·84
Feeding of chicken for 1st month 0.5fr. 4·8d.			
2nd „ 0.6 5·76			
3rd „ 0.8 7·68	1.9	1	6·24
40,000 kilos (88,169lb.) of litter (straw or peat-moss) at 40fr. per 1,000 kilos. (= £1 12s.) is 1,600fr. (= £64) ÷ 30,000 .	0.054	0	0·5184
30,000 kilos coal (66,136lb.) at 24fr. per 1,000 kilos (= 19s. 2·4d.) = 720fr. (= £28 16s.) ÷ 30,000	0.024	0	0·2304
Electricity, 300 fr. (=£12) ÷ 30,000 . .	0.010	0	0·0960
Petrol: One incubator burning for twenty-one days monthly = 252 days yearly, at 4 litres (3·5qts.) per day = 1,008 litres (220.5gal.); fourteen incubators are used = 14,112 litres (= 3,087gal.) per annum. Fourteen rearers burned daily for 320 days, an average of 20·160 litres petrol, has added per bird	0.183	0	1·7568
Percentage allowed to men: Suppose mortality not greater than 2 per cent. (three men each receive 0.03fr. = 0·288d.) .	0.09	0	0·8640
Manager receives per bird	0.04	0	0·3840
Each chick costs	2.701	2	1·9296

The Search for a Scientific Food

	Francs.	£	s.	d.
30,000 chicks cost . . .	81,030.00	3,241	4	0
Twenty per cent. chicks falling off (our numbers are 30,000 birds living; 6,000 chicks have been lost) at 0.40fr. (= 3·84d.) .	2,400.00	96	0	0
Interest on 100,000fr. capital (= £4,000) at 4 per cent. is . .	4,000.00	160	0	0
Two and a half acres, ground rent and manager's house . .	2,500.00	100	0	0
Depreciation of material . .	2,000.00	80	0	0
Wages of three men at 100fr. (£4) per month each	3,600.00	144	0	0
Wages of manager at 400fr. (£16) per month	4,800.00	192	0	0
	100,330.00	4,013	4	0

RECEIPTS.

	Francs.	£	s.	d.
Sale of 15,000 couples weighing 1.750 kilos (3·858lb.) each (average price, low estimate, 8.50fr.) = 6s. 9 3-5d. per couple . .	127,500.00	5,100	0	0
On 50 per cent. bad eggs, 16,000 are in-fertile, and sold at 0.05fr. (0·48d.)	800.00	32	0	0
Manure sold at 60fr. (= £2 8s.) per month	720.00	28	16	0
Refuse, addles, grass, etc., is enough to feed ten sows, each giving twelve little pigs = 120. Each sold at 12fr. (= 9s. 7 1-5d.) at six or seven weeks old . .	1,440.00	57	12	0
	130,460.00	5,218	8	0

PROFIT.

	Francs.	£	s.	d.
Total receipts	130,460.00	5,218	8	0
„ expenses	100,330.00	4,013	4	0
Net profit . .	30,130.00	1,205	4	0

From which 10,000fr. (= £400) should be taken for ten years to refund initial capital of 100,000fr. (= £4,000), and 20 per cent. dividend should be distributed.

Percentages and Profits

	Francs.	£	s.	d.
Percentage allowed to men : If mortality is 4 per cent. each man receives 0.020fr. per bird (= 0·192d.) . .	0.020	0	0	0·192
Three men receive 0.06fr. on each bird (= 0·576d.)	0.06	0	0	0·576
If mortality were 5 per cent. each man receives 0.015fr. (= 0·144d.) per bird	0.015	0	0	0·144
Three men receive .045fr. (= 0·432d.) per bird	0.045	0	0	0·432
If mortality were 6 per cent., each man receives only 0.01fr. (= 0·096d.) per bird	0.01	0	0	0·096
Three men receive 0.03fr. (= 0·288d.) per bird	0.03	0	0	0·288

1fr. = 9·6d. ; 25fr. = £1 ; 1 kilo = 1,000gr. = 2·285 oz. (avoirdupois) ; 1 litre = 1¾ pints (practically).

CHAPTER IX

CAPITAL REQUIRED FOR STARTING

NOTHING can be started without a more or less considerable investment of capital. The installation of our industry comprises three brooder houses, one incubator room, office, cellars, two sheds, one for litter and the other for utensils. These buildings have to be erected as near each other as possible, as to spread them about on the ground would greatly complicate the work. The incubator room should be built on bricks partly underground. A room for the storing of the eggs, another for the eventual packing of them, and a cellar for petroleum and coal should adjoin the incubator room. Above, there should be a big store room as well as a room for mixing the food, an administrative room, and a reserve room for small utensils. The sheds should be constructed near this building, everything must be at hand. The supervision of the business is easier when the manager or head responsible man lives near the brooder houses. Everything has to pass under his eyes—the removal of the chicks from the incubators and the serving out of food and litter. This constant supervision is very economical in practice. Only under these conditions can such a small staff as three men be sufficient. In my establishment everything was worked by two engines of 30 h.p. and 12 h.p. The cost of them and the machinery will require more capital, but they save time and the labour of one man. I did not reckon this outlay in the accounts given, as one can do without them. In the construction of the buildings a future enlargement of the business has to be taken into account. I produced 30,000 birds yearly, and could easily, on account of this foresight, have produced double with an extra building and the probable enlargement of another.

A capital of £4,000 is required for the rearing of 30,000 chickens for the table; the greater part is utilised in the construction of the buildings and the other part for the current expenses of the business. All goods should be bought wholesale and by contract at the beginning of the season, payable quarterly. The first three months the business brings in no money, as no chickens are to be sold before this age. The first advance of capital for these three months amounts to 20,445.62 frs. (= £817 16s. 5d.), of which details will be found in this article.

Starting Smaller Establishments.

For small undertakings no approximate cost can be given, as it is dependent on the importance or manner of production. Should it be monthly, three brooder houses would be necessary and a place for the incubator would probably be found on the spot. Should the production be quarterly a bigger incubator room, as well as a big building, would be necessary. A business on a small scale can very

Capital Required for Starting

often be started with buildings already existing. Some alterations, such as rearrangement of doors or windows or the erection of a wall, would have to take place, but this is easy to arrange. A small capital would be sufficient for the setting up of the business. As for the current outlay, here again it depends on the number of birds produced monthly or quarterly. The details of the current expenses herein given will make the calculations easy. It goes without saying that it will not be so profitable as a big undertaking. Like a large restaurant the more people it feeds the greater profit. The same reasoning applies to chicken manufacturing. This same rule applies also to the heating, lighting and general expenses, as they do not increase in proportion to the profit obtained by a greater number of birds. A small business can never be worked out on the same principle of economy as a large one, where the erection of the buildings and everything beforehand had been calculated with a view to the saving of labour and time. Nevertheless it is a profitable occupation to make £250 yearly out of 8,000 unfattened birds. The difficulty is to start practically and economically.

Need of Expert Advice.

No business should ever be undertaken without previous knowledge, and the counsel and advice of a genuine expert mean very much in economy. The expense of securing a competent expert at this sort of business is a trifle when compared with the saving it may bring in construction and in general management. Many people, unfortunately for themselves, save on the small things and are prodigal in large ones. This is not real economy, especially in this particular case, as this rearing of table birds is totally different from the English rearing. The deficiency of thorough and adequate knowledge can be partly counteracted by the experience of a reliable expert, who, being on the spot, will make the best of everything, teaching, helping in the erection of the buildings, and starting the business if necessary. Advice by correspondence cannot be profitable in this case. Being responsible for the introduction of this method in England, I am ready to place my practical experience at the disposal of my readers either for small or large undertakings.

COST OF BUILDINGS FOR A YEARLY PRODUCTION OF 30,000 TABLE BIRDS.

	£	s.	d.
Brooder house No. 1 (120ft. long by 30ft. wide) .	250	0	0
Brooder house No. 2 (150ft. long by 30ft. wide) .	400	0	0
Brooder house No. 3 (180 ft. long by 30ft. wide) .	600	0	0
16 incubators at £10 each	160	0	0
Building of the incubator room (two floors), sheds for litter and utensils, etc.	1,000	0	0
Heating apparatus, installation of electricity, water pipes, and hydrant, hose installed in all buildings in case of fire	300	0	0
28 rearers for brooder house Nos. 1 and 2 at £8 each	224	0	0
28 strong slatted sleeping boards for brooder house No. 3 at £1 each	28	0	0
Wire netting for double runs, movable partitions, doors of same, numerous food troughs, drinking troughs, scratching boxes, hoppers for the three houses	150	0	0
Boxes for food, buckets, wheelbarrows, utensils .	50	0	0
	3,162	0	0
Cash for current expenses	838	0	0
	£4,000	0	0

Capital Required for Starting

The current expenses of the first three months are as follows :—

	Francs.
6,000 eggs would have to be bought monthly during four months : equals 24,000 eggs at 0.20frs.	4,800.00
The feeding of 3,000 chicks during one month costs 0.50frs. each	1,500.00
The feeding of 3,000 chicks during two months costs 1.10frs. each	3,300.00
The feeding of 3,000 chicks during three months costs 1.90frs. each	5,700.00
Wages for manager and three workmen quarterly	2,100.00
Percentage allowed on 3,000 chickens at 0.13frs. each bird to manager and workmen	390.00
Cost of petrol quarterly	1,375.62
Cost of litter, coal and electricity, quarterly	655.00
Three months' rent of dwelling house, rates and taxes being included in the rent	625.00
	20,445.62

Equals, in English money, £817 16s. 5d.

At this juncture an average of 2,500 birds are sold and brooder house No. 3 is empty. Every following month the same number of birds will be ready for sale. The three brooder houses should be thoroughly cleaned and the birds of No. 2, being two months old, should be transferred to house No. 3. The little ones of No. 1 should in their turn be carried to house No. 2, and house No. 1 should be ready, after thorough cleaning and disinfection, to receive the baby chicks. The incubators are in the meantime hatching their fourth batch of eggs to fill up No. 1 brooder, and so on every month.

CHAPTER X

HENS AND EGG PRODUCTION

THE article on the rearing of chickens would be incomplete if the production of eggs did not follow. I intend to show how the production of eggs and of table birds is obtained in France, my native country. Belgian and French people are very clever with poultry, and did not make the mistake of expecting birds specialised in laying records to give them their table production. They knew it to be useless. But they did better. They made all their breeds good layers, and the newly imported American birds bred exclusively for laying are kept apart on account of their inferior table qualities. Birds so specialised are useless for the table. The same is true of exhibition birds. No double purpose can be expected from them.

In France all ordinary common hens—called in England barn-door—are, in fact, more or less mongrel birds, but good for eggs as well as for the table. France has had for centuries her special way of making birds suitable for the satisfaction of the gourmet. For the benefit of my readers I will indicate the French breeds that would be suitable for this country. Should these breeds, as well as the good English breeds for the table, be selected for laying, as the Leghorn or Wyandottes have been, they would fulfil the ideal of double purpose utility birds. Do not tell me that the English climate is not so suitable, etc. I know all the objections that can be raised. Birds that could not be bred profitably in England are indeed very few. It will depend on the way they are kept. For this reason, I give in the following list of birds some particulars about them. The productive American birds have to be kept also in a way suitable to them. The same should be done with French birds ; but unfortunately this is not done. I have seen the delicate Houdan and other French breeds kept in the damp climate of England quite as if they were natives of the land. This is absolutely different to the way they are kept in France. Each breed requires special soil or adequate care to enable it to live out of its country and be profitable. Otherwise it does not get acclimatised, or if it does physically, it is with detrimental effect on its productiveness. The laying strain and the laying competition birds are kept in a more or less artificial way. Their conditions of life are different to those they would find were they living freely as Nature meant them to live. It is plain that when intensive production is aimed at, lodging, foods and management should be different. Some useful hints will be given in this new series of articles about these requirements.

The reason why the production of eggs in England is chiefly obtained from birds of American importation is owing to the specialisation made by the American breeder in laying strains. There is also the desire for rapid profit without much trouble.

Clever English Breeders

These reasons are helped by a boom in all American innovations, in many cases well deserved.

English Breeds.

English people are very clever breeders, and perfectly able to do as well as American and French people. Why they do not take the trouble to select and build up good laying strains from the splendid table birds they possess is difficult to understand. The Sussex, old Kent, Surrey and Dorking fowls could easily be birds of dual purpose. It would surely pay to study them for this purpose, because the chickens obtained from such breeds, being of better quality for table, would obtain a higher price on the market than those coming from inferior birds. They should easily make good the possible deficiency of eggs during the first years by the better price that people would obtain for their surplus birds and cockerels during the time necessary to build up a laying line.

France is more than self-sufficing in eggs. It is a well-known fact in this country that the quality of her eggs and table fowls is first class. France exports not only to England, but enormously to her colonies. Exact figures of her production are difficult to give, as many of the French eggs pass the frontier and are bought by the neighbouring countries without appearing in her markets' statistics. Ten years ago the export from France to England amounted to 3,000,000 eggs. At that time England imported 1,250,000,000 eggs, for which £4,000,000 were paid. At the same time the poultry statistics

Hens and Egg Production

of France gave the astonishing number of 45,000,000 hens, and from information just sent to me from the Board of Agriculture the number is 632,741 hundreds of eggs for 1914. The American breeds have also obtained a footing in France, but cannot have such a bad influence in France as in England, because France possesses an enormous number of good layers, also good as table fowls. The national commercial sense will not permit these American birds to multiply to any great extent and flood the markets with inferior birds. Anyhow, the French, being connoisseurs of poultry, would not accept as roasting chickens such an inferior quality of birds. The special French production of eggs comes from their splendid birds, nearly all of them as good for table as for laying. What France can do, England is also capable of doing, when she understands that a bird capable for double purposes can be obtained with skill and patience. So far her double purpose birds have been exhibition birds allied with pretended utility. Both qualities are very seldom found together in English breeds.

All the breeds of France possessing the above qualities are fond of liberty and activity, but a number of them accustom themselves to any surroundings, and also to captivity without losing their productiveness.

The following instructive enumeration will enable the English breeders to indulge their fancy this time with judicious crossings and to produce good layers and good table birds. I am convinced that as soon as they have recognised the quality of such birds, the size they reach and the profit that can be obtained

Laying Power of Various Breeds

both from their laying power and from their table quality, they will not breed any more from inferior birds. Such conservatism should have changed with the war and the new conditions of life, surely! The birds in the following list are common barn-door fowls. Some of them are not suitable for England under ordinary conditions, but those will be pointed out. All free range hens will be denominated *Rustic*. The number and the weight of eggs quoted are an average. None of these birds has ever been trap-nested.

BERRY HEN—200 eggs, weight 60grm. Splendid table chicken easy to fatten.

BOURBONNAISE—170 eggs, weight 65grm. Coloured shell.

BOURBOURG—170 eggs, weight from 65grm. to 70grm. Salmon coloured shell, precocious chicks and very delicate flesh.

BRESSE—175 eggs, weight 67grm. Flesh extremely white and juicy, renowned for ready acceptance of forced fattening.

COURRIERES—Cross bred from Bresse-Langshan, very big. 200 eggs, easy rearing, enormous pullets.

CREVE-CŒUR—160 eggs, 65grm. to 70grm. Very delicate flesh, remarkably white, unsuitable for England, dislikes wind and damp. Would do for intensive or semi-intensive culture.

BARBEZIEN—160 eggs, very big, from 75grm. to 85grm. Exquisite flesh, good chicken for Flemish rearing.

CAUSSADE—Rustic. 140 to 180 eggs. Splendid little pullet, delicate and fine.

COUCOU DE RENNES—160 eggs, 60grm. to 65grm. Poulet de grain 2 kilos at four months, without forcing; very delicate flesh.

FAVEROLLES—Salmon type, Barn-door in France, 150 to 160 tinted eggs. Very precocious and splendid quality for the table.

LA FLECHE—Enormous hen, exquisite flesh, unsuitable for England. Chickens rather difficult to rear in liberty, splendid for Flemish rearing. 150 eggs, weighing from 70grm. to 76grm. Suitable for intensive culture.

CAUMONT—170 eggs, weighing from 60grm. to 65grm. Rustic.

COURTE-PATTE—150 eggs, weight 65grm. Requires dry soil; would suit intensive system; good chickens.

Hens and Egg Production

GATINAISE—Rustic, 180 eggs, weight 75grm. Delicate chicken ; flesh extremely white.

GAULOISE—Rustic. 150 eggs, weighing from 50grm. to 55grm. Most suitable for the production of Petit-Poussin, these being always superior if they come from small birds, because they are matured.

GOURNAY—Extremely rustic, fond of living in trees. 150 to 170 eggs, weighing from 60grm. to 70grm. Very delicate flesh.

HOUDAN—Splendid for both purposes, unless it has passed into the fanciers' hands. Unsuitable for England, must be reared in confinement, dislikes damp and wind.

MANS—130 eggs, very big, weight 70grm. to 75grm. One of the most renowned (famous Poulardes du Mans). Unsuitable for England, dislikes damp, would suit intensive culture.

GASCONNE—150 to 180 eggs, 60grm. Lays at 4½ months or 5 months in winter. Chickens quite ready for table at 3 months, without fattening.

ARDENNES—Rustic. Good brooder. 140 to 160 eggs. Delicate flesh, first class little chickens.

MANTES—Rustic. Lays in winter 180 eggs of 60grm. Pullet extremely fine, good for fattening at three months.

PARVILLY—Very rustic, but slow grower, lays 250 eggs. White shelled, no tendency to brood, splendid chickens.

ESTAIRES—150 to 170 salmon coloured eggs, succulent chickens, precocious chicks, most easy to rear.

HERGNIES HEN—Mongrel Barn-door. 160 to 180 eggs, from 60grm. to 65grm. Splendid for table, very rustic.

French people have been most careful to keep the quality of their birds apart from the fanciers'. People going in for exhibition do not consider utility production. They sell the eggs for exhibition purposes. The peasants have never practised injudicious crossing and recrossing for the sake of conventional beauty of no productive value. We shall see later on how clever the French peasants are in detecting the good from the bad layers. Were those above-mentioned breeds worked as cleverly as the American breeds have been for laying, they would be surpassed by none. Were I to live in England, I would be quite satisfied with the original birds, but would improve their laying quality, and rear in the meantime one of the above French layers. I never recommend the Coucou de Malines, although I myself breed from it, the Belgian market in this case being my master. This bird is a bad layer and possesses big bones. It would not satisfy the English gourmets as much as their own Sussex fowls. The Coucou de Malines has been badly handled by the Flemish peasants. It has become rather delicate, and here has lost its flesh quality for some so-called improvements imposed on it by English fanciers. The same is true of the French Faverolles, Campines and Braekels, all utility birds, which have become here exhibition birds, the two last very good layers of big eggs.

A proverb in France says, "Who runs after two hares loses both." In this case only one is lost. Unfortunately, in this competition, it is utility that disappears. To find out the good layers from the bad ones the old and clever breeders of France have recourse to several tests. They consider the width of the hind quarters of the hens. They have learnt from observation that there is a connection between the hind quarters and productiveness. It is a well-recognised fact in cattle. The same holds good in birds. I do not mean the depth of the hind quarters. This, being due to a relaxation of the abdominal muscles, has nothing to do with laying power. On the contrary, frequently such hens have a tendency to be egg-bound. On account of the looseness of the ovarian organ an egg may fall

Hens and Egg Production

in the hanging oviduct and be difficult of expulsion, whereas when the bones of the pelvis (side bones easily felt in the rectum) and the legs are well set apart, it indicates a better development of the abdomen, consequently of the ovarian organ, leading to a greater tendency to lay well.

Some breeders test their hens with their hands as soon as the hen is laying, which is easily detected by the redness of her crest and her general liveliness, or by trap-nest. They insert the knuckles of the two first fingers into the rectum. All good layers should permit an easy introduction of the knuckles; should the passage be found too narrow, it will show that the bones of the pelvis are too near each other. The hen should be discarded, as she will not pay for her food. The more easily the passage yields to the fingers, the better the laying capacity of the hen. As for the young pullets, as soon as they begin to lay, they should also be tested and the passage should be developed enough to permit the test; if not, a future development would not enlarge the passage sufficiently. These birds will never be good layers. I have applied this method personally for years to my birds and have ringed and set apart the best selected birds. In another pen I placed the medium good ones and, in a third pen, the thoroughly bad ones. Hardly ever had I to alter the first selection. Many other ways are said to be good for the detection of good layers. I shall not enumerate all the methods practised everywhere, such as, the first one to come out in the morning, the last one to go to bed; those that have laid the first in the season and in all weathers, etc., because they are not easy of application. People in a commercial undertaking cannot be there morning, noon, and night, for studying and selecting their birds. Time is only to be spent when there is a certainty of a direct profit.

The laying Belgian breeds—Campines, Braekels, Brabançonnes, etc.—are selected by the sire. I am perfectly well aware that in England this practice is also applied, but differently. A cock is selected, son of a good layer, but hereditary quality does not always answer. We know it by " Reversion to Ancestors " (Atavism). It often happens that a bad bird may come from a good stock and vice-versa. I believe more in visible and tangible qualities. I am very much opposed to line-breeding (in-breeding) for practical laying purposes. It is very good to intensify some quality—no excellence would ever be attained without it—but unless it is very carefully handled it is a dangerous process. It may alter the stamina of the breed, which principal object will be, in our case, rusticity, and consequently diminish the laying power. In the meantime, in-breeding is only a tentative method. Even with the mating of the same or nearly related blood, nothing is certain concerning the offspring. I would use it myself for the fixing of an external quality, as type-plumage, etc., but not for laying. The following method will answer splendidly without altering the vitality of the birds, especially when this method is allied with the above French testing, as it will increase vitality, size and production.

CHAPTER XI

"BELLE ORPIGNE" PHOSPHATE MIXTURE

I WILL not delay giving the recipe for the phosphate mixture any longer on account of the moulting season that has already begun, as it will help the hens to cast off their feathers easily and renew them quickly. This mixture will be found most useful for the rearing of table birds and for the laying season; it eradicates all leg weakness, cramp, etc. It would have been given before had I not been trying some chemical adjuncts to it with a view to keeping the liquid mixture in a pure and sweet state longer than it generally does. Hydrochloric acid or chloroform would, under certain conditions, have answered very well, but these chemical substances might be dangerous, and the remedy worse than the evil. Up till now I have not found anything that would not affect the birds while acting efficaciously as a preservative of the mixture; therefore we shall have to put up with the necessity of making this concoction fresh every day. We must never lose sight of the fact that the liquid mixture ought to be used absolutely fresh, otherwise it might scour the birds and lose its effect.

When my readers have tested it for some time they will realise the practical worth of my advice, but it must be remembered that a fair test can only be made if all the instructions given are faithfully carried out. The proportion must be no less, no more, than according to the directions; no sudden increase or decrease. Contradictory results may easily be obtained from the same mixture when used by different persons, simply because of the difference between careful and careless following of the directions. How often people say when speaking of drugs, " If I give a little more it will not hurt, it is harmless." It might hurt all the same ; consequently no fanciful interpretation of the instructions should be made, but they should be followed carefully. Let me give you an example of the different effects that may be obtained according to the difference of application.

Supposing this phosphate mixture is given longer than is indicated to table birds, the result would be that they would lose their table qualities as they would be too active. We want, in the rearing of table fowls, to see the birds grow quickly, but, once they are of sufficient size our object is to make them put on fat quickly and easily, and keep the juicy flesh quality of the young birds. The length of time this mixture can be given has been tested, and must be followed exactly if the best results are to be obtained.

The Recipe.

" Belle Orpigne " Mixture : Take six tablespoonfuls of bran, six tablespoonfuls of black oats with the husk, four tablespoonfuls of Egyptian lentils, six tablespoonfuls of rice with husk, three tablespoonfuls

" Belle Orpigne " Phosphate Mixture

of big yellow maize. Put these twenty-five tablespoonfuls of cereals in a big enamelled vessel, pour
over it six French litres of cold, hard water (a French
litre measure can easily be obtained in London).
Place the vessel on the fire and let it boil without
cover for three hours. A third of the liquid will
be absorbed by the grain and two litres evaporated
by the boiling, and two of extract will be obtained.
The whole of the mixture should be strained through
a medium wire sieve; the liquid obtained will be
yellowish and of a mucilaginous texture. It should
be used warm, mixed in the first mash given in the
morning. This quantity of liquid will at first be
sufficient for twenty-five full-grown birds till they
are accustomed to it. Part of the beneficial effect
that may be anticipated has already been explained
in the Principles of Feeding articles.

Nature and Effects of the Food.

We know that the phosphates extracted from the
soil are not retained in the body but are eliminated.
The celebrated scientist Boussingault wrote: " The
phosphates to be absorbed by animal or human
organisms must first be elaborated through a vegetable vehicle." This scientific fact has since been
recognised by Hutchinson, Roberts, Kellner, Voigt,
Schultz, Liekernick, etc. From the above mixture
we also obtain a large amount of lecithin. This
phosphoric substance may be regarded as an accu-
mulator of energy.

According to Schultz and Liekernick the vegetable lecithin has the same chemical composition as
the animal lecithin of the egg. It is found in quanti-

ties of seeds, plants, spores in young growth, in cereals, etc. This substance plays a most important part in the body. It forms bones, nerves, capillary system and feathers; it is found also in marrow, in brain, in fact all through the living body. Potassium, which is present also in a large proportion in this mixture, has a great influence on growth. According to Professor Dehérain, potassium has a direct connection with the development and vigour of the animal or human being. The mixture also contains manganese, magnesia, calcium, oxide of iron in varying proportions. All these salts are acted upon in the stomach by the gastric juices and bacteria. They vitalise animal energy by liberating warmth and electricity in the cells, which immediately absorb them. This mixture is a strongly condensed addition to the insufficient amount the birds obtain from their food.

The absorption of the same amount of mineral and vegetable salts is an absolute impossibility, on account of the enormous amount of food that would have to be eaten by the birds. It replaces the commercial phosphates given in the Flemish ration. It costs very little compared with the price of the commercial phosphates and acts differently, being more efficacious and productive. The beneficial effect on health and growth is wonderful. I gave it to my birds for years, till I found a concentrated way of administering it in a dry form. This was more convenient for a paying business, as owing to the negligence of my assistants, I had experienced diarrhœa in the birds, when the mixture had not been freshly made. The introduction into this

" Belle Orpigne " Phosphate Mixture

phosphate of several items rather difficult to handle
on account of their varied properties, which have to
be carefully measured and analysed, makes its manu-
facture impossible for people who do not possess
sufficient chemical knowledge and special machinery.
It is a more complete product, fulfilling every possible
requirement of the birds, and permitting the breeder
to dispense with milk for the rearing of table fowls,
a substitute for milk being included in it. Thanks
to the intelligent utilisation of the waste products
and the extra weight of birds obtained, its cost is
repaid tenfold. In the condensed dry form, the
vehicle chosen for these phosphates and vegetable
and mineral salts, is in the meantime a concentrated
food of high nourishing value, which has to be added
to the ordinary feeding-stuff. This dry concentrated
phosphate will be on the English market in a short
time under the registered trade mark of " Belle
Orpinge." The dry form keeps indefinitely without
losing its quality or power.

How to Use It.

In whatever way it is given it will eradicate
anæmia, paleness of face in laying hens kept in confinement. Its great value will be found in the
stimulation to laying and in the fertility of the
eggs and strength of the germs, thereby producing
greater vitality in baby chicks. The application of
the liquid form should be as follows : For the rearing
of table birds, four litres should be given to 100
chicks from the first day of feeding, and increase
slowly to eight litres till they reach six weeks old.
Then decrease till complete cessation when the chicks

are eight weeks old. The effect continues for some days.

When given to chickens meant for breeding and laying purposes the same course should be followed till they are four months old, and slowly decreased to four litres till the birds reach their fifth month. When given to twenty-five hens, two litres should be given the first week, increased to six in moulting time, then gradually decreased after feathering to four litres, which are to be given all the year round. The mash in which the liquid is introduced should be sparingly given in order to compel them to eat it all up.

The birds are generally very fond of it. The grain has to be cooked—not soaked—not less than three hours in the evening, then covered and left on the stove the whole night. It should be strained in the morning and used immediately. Should the quantity of liquid be too great for the mixing of the mash, the remainder should be given as a first drink in the morning, but it is better used with food. If the mixture should be found cold in the morning, the thick gelatinous cream, which will be found over it, should be skimmed off and mixed in the mash, as it contains appreciable nutrients. Then the liquid should be warmed up, strained after warming, and used immediately.

The by-products of these phosphates are for the greater part carbohydrates. It has a great value as food for pigs, but not for hens, because it is too fattening for laying hens and will produce inferior quality of fat for table fowls. It should be consequently utilised for pigs. This waste product

"Belle Orpigne" Phosphate Mixture

might be eventually easy to sell in the neighbourhood for the purpose of fattening pigs.

Hens and Egg Production.

To believe in science, as far as theory is confirmed by practice, does not prevent believing still more in the deductions and observations made by simple people. How many times learned persons have closed their eyes to facts that, according to science, were improbable, until they had to realise that science is not infallible and can be advantageously assisted by practice. Closer investigations of apparently insignificant factors believed to be of no importance had led scientists later on to recognise that the common-sense of peasants was superior to their theories. This good practical sense did not show itself in Belgium, where the production of eggs for table birds is obtained. For the last twenty years the enormous development of our rearing industry has roused in the Flemish peasantry an avaricious spirit which has made them kill " La poule aux œufs d'or." Their ancient skill in the handling of birds had totally disappeared in view of immediate profit. Nothing else appealed to them except that the more eggs they could obtain the more they would have to sell, or the more birds they would have to rear for table. They never thought of the future. So it was that, after the first year of laying, they got rid of the young hens in order to avoid feeding them during the winter. Only the young immature stock was kept for laying and breeding purposes. Nowhere has a too close relationship been avoided. Pedigree was not even thought of, and inbreeding was practised everywhere indiscriminately, the consequence being that the laying power and the stamina of the national breed has deteriorated. The standard of laying has decreased to such an extent that this bird has become a very bad layer and is kept only for the production of table fowls. Fortunately, the utility breeders of the Campines, Braekels, and Brabançonnes have managed better, and kept up the stamina of their stock. These breeds are excellent layers, the last named very good for table, also very rustic, laying an average of 170 eggs per annum, weighing 70grm.

Selection of the Male.

To preserve the productive quality of these birds the breeders select their hens not only for general liveliness, activity and parentage, but they do not neglect the important factor of the cock. They have noticed that the daughters of their selected male develop the quality of good layers. The cocks that crow first, those showing a precocious tenderness towards the hens, those demonstrating a highly combative instinct are thought to be of greater vitality and able to influence favourably the fecundity of the offspring. These birds are reserved for the annual mating. Practice and science have proved that such precocity of sexual and combative instincts correspond in the female offspring to a greater precocity and prolificness. Therefore good layers must not be spoilt, or the laying powers of their offspring lessened, by an injudicious mating. This way of selecting indicates that those breeders, with great common-sense, rely on heredity as far as the

" Belle Orpigne " Phosphate Mixture

vitality and stamina of their breeding birds are proved by observation.

Cock-Crowing Competitions.

Another curious method of selecting the cocks is applied by the Belgian breeders of the small and beautiful bantam breeds ; the Ardennais, the Bassettes and the little bearded bantam of Antwerp are selected by a crowing competition. Their breeders come with their little cocks enclosed in baskets. These are numbered and placed on a long table, side by side. Each bird remains in his basket. The people sit about, and the judges, watch in hand, are ready to take notes. Every crow is registered near the number of the crower. Big bets are staked on the birds as well as big prizes. The bird that crows most times in the hour is the winner of the competition. He is considered to be the most pugnacious and precocious of all the birds present. He is the idol of the house, as he not only earns money by his vocal powers, but he is used for breeding, the eggs coming from his mating pen being in great demand as he improves the laying power of the hens. The buyers expect still more notable fowls from the eggs bought for incubating purposes.

This way of selection combines pleasure and profit. It improves the utility quality of the birds, procures innocent amusement, and satisfies the gaming instinct that exists in so many people. It is in the meantime deprived of the cruelty of cock-fighting. Nothing prevents this selection being made from bigger breeds. The birds subjected to a crowing competition develop a desire to be first. It increases their vigour by the rage it incites in them at hearing the other cocks crowing, and it also stimulates virile qualities.

The Utility of the Trap-nest.

Another way of selecting hens for utility purposes, is by the use of trap-nests. There exist several varieties of them working on the same principle. The hen enters the nest and by so doing unlocks the door which closes behind her. When she is shut in no other hen can enter the nest, and she herself cannot get out unless somebody releases her. One of the drawbacks of these nests is that, if the hen is not released quickly enough, she becomes nervous and indignant at being a prisoner, and she tries by every means to get out, with the result that her egg may be pushed against the wall and get broken. As soon as the hen is attracted by the moisture of the broken egg she tastes it and likes it, and straightway becomes an egg-eater. Such hens are so clever that they leave absolutely no trace of their mischief behind them. Once this bad habit is started it is very difficult to cure it. Numerous trials have been made but none has proved reliable. Sometimes it also happens that a peaceful hen finding herself perfectly happy in the quiet of the trap-nest would remain in it for some hours, half asleep, preventing her comrades from occupying the nest. Ordinary trap-nests have to be numerous in order to be practical, and unfortunately they entail the obligation of releasing the hens several times in the day in order to identify the eggs, this being absolutely necessary when birds are bred for pedigree layers or exhibition

" Belle Orpigne " Phosphate Mixture

purposes. In a commercial undertaking we have to
select our hens in a simpler way, because this liberating is costly in time and attendance. Wherever the
hens are numerous, a man will be occupied a great
part of the day in releasing them, and as all assistants are not to be relied upon, it is better to do without this risk and expense.

CHAPTER XII

TRAP-NESTING ECONOMY

THE collecting of eggs from trap-nests offers a
temptation that it is better to avoid. The disappearance of ten eggs daily in a pen of fifty birds
makes a total of 300 eggs lost in a month. This
repeated in several pens means a loss in the business.
Trap-nests are used specially for pedigree layers, for
the breeding of exhibition birds and for laying competitions. For utility purposes they are unfortunately not so much used as they should be, on account
of the drawbacks mentioned in the preceding chapter.
In the production of eggs either for market or for
table birds, where some hundreds of birds are concerned, the French and Belgian peasant methods are
difficult of application, but if they are combined
with trap-nesting a maximum of production will be
obtained without altering the stamina of the birds.

Practical trap-nesting should be used in a way to
suppress the expense of one or several attendants'
loss of time and the eventual disappearance of eggs.
In the commercial production of eggs it is not
absolutely necessary to identify the hens with the
eggs laid. A nest permitting one to detect the
layers from the non-layers, and from which the hens
can liberate themselves and set it again ready for

Trap-Nesting Economy

the ones following, should be sufficient. But with such nests the layers have to be set apart from the pen in an enclosure, which reduces the floor space. That is the reason, as I was often told, why these nests are not in vogue. It is easy enough surely to pen the hens without interfering with floor space. I, for my part, know of several ways of doing it. In the opposition to more modern methods lies the illogical conservatism that exists so much in this country. There is always a way to succeed if only one has the will to do it, but up till now this sort of nest has been declared unpractical. Is it better, or more practical, as I have so often seen, to have nests so constructed that they allow several hens to sit together? Not only do these nests give no indication of the laying power of the birds, but they increase the danger of breakage and easily start the habit of egg eating.

I quite agree that the farmer takes no notice of the barn-door fowl's fertility. These birds finding the greater part of their nourishment for themselves, he does not attach much importance to the fact whether they are good layers or not. Where poultry is kept for profit trap-nests are necessary. We have to increase their laying power by selection and remove the useless birds that are increasing the feeding expenses and raising the price of the eggs produced by the industrious ones. Some specialists in egg production declare themselves perfectly satisfied if they obtain from twenty-five to thirty eggs daily from pens of fifty birds. There is no doubt that in every pen there exist several worthless birds, and why should they be kept if it is proved that they do not pay for their keep? A just objection people raise against trap-nests is the cost and risk of attendants necessitated for liberating the hens from them. On the other hand, being averse to the special enclosure necessary for the automatically liberating trap-nests, they remain *in statu quo*, and do not obtain better birds because they do not select them. This is very well for the backyard holders who attend to their birds personally and who, owing to the small number they keep, may know their individual characteristics, either by observation or by the application of the peasant methods.

Trap-nests necessary to Intensive Production.

Whenever intensive production is practised it is worth while to get hold of good automatic safety trap-nests. Unfortunately many people who do not mind paying a good price for poultry houses, incubators, rearers, etc., draw back at the expense of a good and reliable nest which would place the eggs out of reach of the hens' beaks and of indiscreet hands. These nests would obviate the risks and the necessity for attendants, and repay the initial outlay in a few weeks. A serviceable and self-liberating safety nest evidently costs more than a set of ordinary trap-nests. A trap-nest, to be satisfactory, should prevent breakages as well as several hens sitting upon the eggs, this being liable to spoil the freshness of the eggs, and consequently nothing would be better than the use of a trap-nest, so conceived that the eggs would gently disappear from the nest into a double bottom as soon as laid.

Trap-Nesting Economy

Once the eggs are placed in security, their removal should take place only once a day, either in summer or in winter. In this case the nest should be padded in such a way as to prevent the cold affecting the eggs.

I will show in the following chapter, " Different methods of Housing Poultry," how I managed years ago, partly without assistants, who were allowed to enter the pens only after the birds had gone to roost and the eggs had been removed and identified. At the time I was breeding according to standard, studying in the meantime in some other pens the way to improve egg production. I was also testing the French and Belgian methods of detecting good layers in conjunction with trap-nests.

These practices, as well as trap-nesting, have been of great benefit to me in every way. There has been a gain of precious time, many more eggs, and fewer people employed. The work in the poultry yard in the daytime consisted of feeding and watering. The rest of the work was done very quickly, as we shall see later on, and the remainder of the day my assistants were occupied elsewhere in repairing, gardening, or carpentering. After the birds had gone to roost the litter was attended to, shaken and cleaned, so that it could be well ventilated and dried in the course of the night.

By the way, the management of the birds in my system of housing is totally different to the methods employed in this country. Wherever trap-nesting is applied a laying record is kept. In each pen a sheet is hung, to show how many eggs are laid by each bird. The frequent repetition of a sign at the date of the day opposite the number of the hens indicates the best, and distinguishes the good ones from the medium or bad ones. It has been said above that the self-liberating safety nests can be used without the intervention of attendants. A special simple device renders them, if desired, specially useful for the fancy and pedigree breeder for the building up of a laying strain and the automatic control of the laying competitions, as it identifies every egg laid, with the hens that have produced them, with absolute precision. It detects also the hens that have sat on the nest without laying.

Big Eggs or Heavy ?

Some breeders attach great importance to the production of big eggs. I do not see the necessity of breeding specially for big eggs so long as the eggs are sold by quantity, because there is no profit in it. Breeders of big eggs lose sight of the fact that the obtaining of them means a greater absorption of nourishment, which is not compensated for by the slight difference in price obtained. Should big eggs for the market be thought necessary it would be more practicable and less costly to breed direct from hens renowned for this constitutional characteristic. As for incubation purposes, it is erroneous to think that bigger chickens would be obtained from bigger eggs, because they have more room to develop in the shell, etc. Many experiments made on this subject by myself have taught me that greater importance should be attached to the weight of the eggs coming from hens of the same breed rather than

Trap-Nesting Economy

to the size, on account of the density of the contents. I know without doubt of the superiority of density over size because my experiments have always been made without the slightest help from assistants. Besides, nobody can deny that Cochins, Brahmas, Coucou de Malines, Plymouth Rocks, etc., lay very small eggs in proportion to their size, but from which, nevertheless, enormous birds are bred. This is good evidence that the size of the eggs is not so important as it is thought. On this account trap-nesting for big eggs should be left alone and utilised more for the improvement of quality, prevention of egg eating, etc. Several nests are designed to prevent this trouble, but few are really practical, requiring, like the ordinary trap-nest, constant attention.

CHAPTER XIII

TRAP-NESTS TO PREVENT EGG EATING

TRAP-NESTS for preventing egg eating have serious drawbacks. For example, a hole is provided in the middle of the nest through which, when an egg is laid, it disappears, or is removed by means of a sliding board leading to another one which directs the egg into a conical double bottom where the eggs are stored. But after the first egg has reached the bottom it is liable to be cracked by the second egg rolling against it. A newly laid egg is soft and pliable, but the calcareous matter of the shell quickly solidifies by exposure to the air. The hen does not always leave the nest immediately after laying, but may remain seated there a little while. When she rises the egg is quite hard and causes damage by rolling against the preceding ones.

To obviate these breakages the removal of the first egg as soon as it is laid is necessary, entailing constant attention and equivalent loss of time. Again, these nests are dangerous and may injure the hen which, in going in or out, may put her legs in the hole or between the sliding boards. Hens dislike the feeling of a hole underneath them. They do not feel secure over it, nor do they like laying on the sliding boards on account of their instability. They

Trap-Nests to Prevent Egg Eating

much prefer their old shabby nests or laying on the ground. Some other means, therefore, had to be thought of for the removal of eggs, and mechanical trap nests began to appear. Certain of them are marvels of mechanism, but so complicated that they easily get out of order. Anyhow their appearance puzzles the birds immensely. The simple creatures understand a nest only when it is firm, deep and soft, with nothing unusual in or around it. They also dislike being trapped in tiny cages after laying, in order to have their eggs identified. In such little enclosures the poor things are longing to rejoin their companions. They very soon associate these ideas together; nests mean imprisoning. Consequently they discard these nests altogether and lay where they choose.

Various Methods of Trap-nesting.

Some other sort of trap-nests for preventing egg eating have not only a hole in the middle of the nest, but a delicate and very sensitive trap or india-rubber pad through which the egg disappears. The newly laid egg being meant to work the mechanism by its own weight, this causes the trap or india-rubber pad to open and, in the meantime, to act on a catch which releases the closed exit door. Imagine the hens treading on this sensitive device and you realise how impracticable it is. Moreover, as the weight of one egg may differ greatly from another, the mechanism has to be arranged for the working of a very light one, to yield on very slight pressure. Some other nests have ringed holes in the middle of the exit door through which the hens are meant to put their heads and to push in order to open them. In going away they carry the ring on their necks with them, What do the poor birds think of this, I wonder, when I see them doing their best to rid themselves of such feminine adornments. These rings being meant to aid in the identification of their eggs, they have for this purpose to be caught and de-

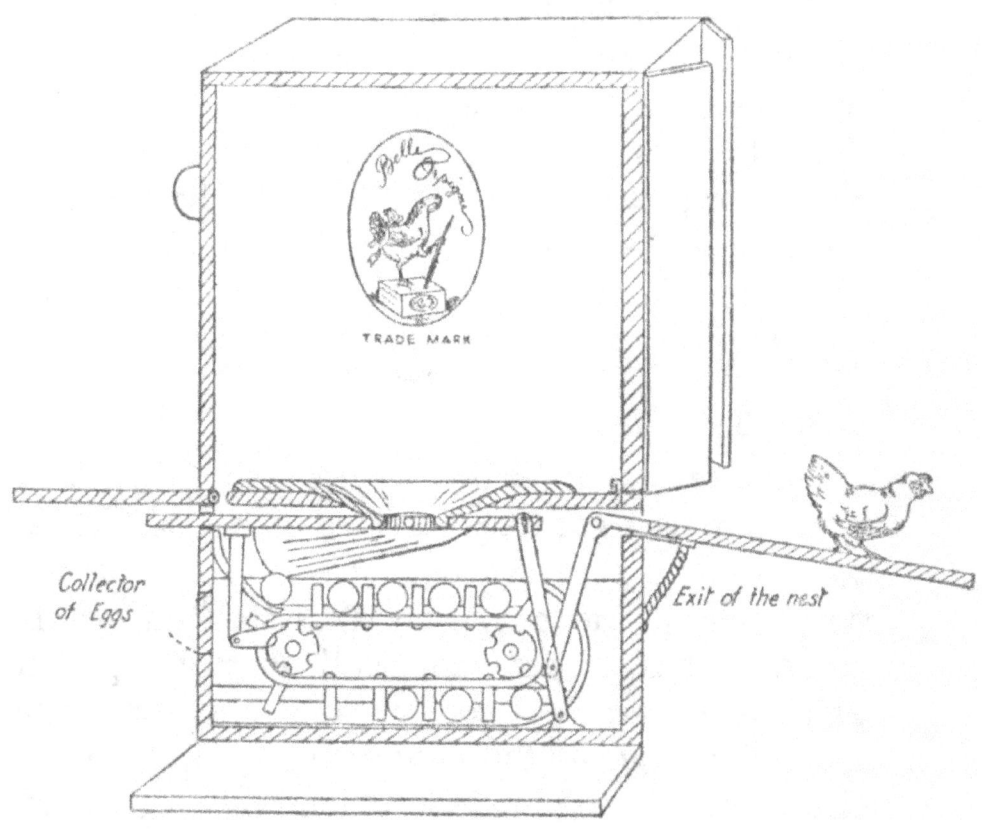

Collector of Eggs

Exit of the nest

TRADE MARK

prived of their necklaces, either by chasing them in the daytime or delivering them up at night.

Some other nests are constructed with different exits or with interposition of curtains. Last, but not least, I know of one nest that paints the hens with colour to identify them when they leave the nest. What becomes of the paint when they go into another coloured nest the next day I do not know. This

Trap-Nests to Prevent Egg Eating

macedoine of colouring has certainly a lively effect in the yard, but its practicability escapes me. I was very amused in going through this collection; the only one I did not find among these fancy nests was a musical one. Joking aside, although I had sometimes to admire the mechanism of some of these apparatus, I wondered how it was possible to expect the hens to make good use of these nests without having them undergo a course of instruction? The nests invented are innumerable, which shows that a practical nest, working to give every security, has been a long-felt want. A simple nest having only one entrance and one exit, that appears like an ordinary nest, and whose mechanism cannot get out of order, is the thing that will answer our purpose.

A Practical Trap-nest.

I used several kinds of trap-nests myself for nine years, but soon got tired of them on account of the necessity for perpetual attendance on those of simple type, and of the complicated mechanism of the scientific ones. I gave them all up and began to ponder over means of constructing a poultry house and nest that would fulfil my exacting requirements. These were, firstly, to prevent entrance and interference in the poultry house, where very often eggs are laid on the litter, even when the ordinary nests are used. Secondly, to get accurate records of the eggs laid by each hen, and to have them removed without risk of breakage. Thirdly, to be able to detect the egg-eaters. The innumerable trap-nests I have invented and constructed before finding an automatic safety liberating nest—which succeeded beyond my expectations—would fill a big room. Here again I had proof that nothing can stand against patience, time, determination and love of the undertaking, whether hobby or business. These are irresistible factors towards perfection. At last I found a comfortable and simple trap-nest with nothing in it that would appear suspicious to the hens. It removed the egg, liberated the hen, and identified the layers. Its working is simpler than an ordinary nest, as it is quite automatic.

Belle Orpigne's Harness.

One other little invention of mine completes the particulars of the trap-nest—it avoids the handling of the birds generally practised in the identification of the laying hens. Until now birds have worn aluminium rings on which numbers are stamped. Too often these rings are unpleasant to read on account of their dirty condition through mud or other matter—which have to be scratched out in order to detect the identification number. They wasted time and did not satisfy me. With the help of this Harness birds are identified at a glance; there is no necessity to handle them; they wear this light device between their shoulders without even noticing it. The number is printed in black, big enough to be easily detected. The harness is made in coloured celluloid which indicates at the same time the age of the bird. It avoids the necessity of the celluloid spiral rings generally in use to indicate their age: for example, should the hen wear a red harness it would mean that she is three years of age— a blue one that she was born two years ago—a green harness that she was born this year, etc. Unfortunately it was impossible to obtain harness made of celluloid for the dairy show of 1916. They have had to be shown made of aluminium—as they ought to be very light in colour. In pedigree or laying strain breeding this harness will soon teach the poultry breeders to associate the productive

NOT IN USE   IN USE

quality of the hens in conjunction with the peculiarity of their shape so often indicative of their laying power and other qualities. It will also help to detect vicious hens caught in the act of their mischief without having to chase them—it will enable the breeder to pick out from the pen the hens that he desires to pen away. It is a most useful little device which saves time and labour.

Commercial Rearer Required.

I intend, in the future, to breed my laying strain for the production of eggs necessary for the rearing of table birds. Unfortunately my property and rearing establishments are now in the hands of the

Germans. It is more than doubtful if I return to Belgium that I shall find them as they were. I have to contemplate the necessity of reconstructing trap-nests, poultry houses, etc., because the position in which I shall find myself will prevent me investing such a large amount of capital as I did before. These consequences of the war lead me also to think of the construction of the commercial rearer I was going to build for big undertakings when the war broke out, because this apparatus will avoid the construction of the brooder house No. 1, and in certain undertakings will permit me to do away altogether with the brooder house No. 2.

A single commercial rearer will take care of 800 to 1,600 chickens during a month or half the number during six weeks, and will provide the chicks in both cases with ample accommodation. They will have in it a floor space three times as large as is generally given for the same number of birds in other foster mothers or rearers. In fact, the rearers have more room in a section made for 200 chickens than they used to have for 240 birds in the rearers I used to utilise in my big concrete brooder houses. I had been working out this apparatus since I realised years ago the drawbacks of all the big commercial rearers. (See the third and fourth chapters.) This new commercial rearer is not only easy of attention and cleaning, but also very economical in fuel, and fulfils all the laws of hygiene, ventilation, self regulation, and temperature, features too often neglected or badly understood. It makes it impossible for the chicks to get overheated and does not allow of the accumulation of carbonic acid on the ground. For the production of Asparagus chickens (Petit Poussin) it would be most convenient and handy, as the chicks can live in it easily till market time.

For people raising poultry on a small scale, small rearers for 100 birds as well as outdoor rearers —easily movable and worked absolutely on the same principles—will be constructed in the near future as well as the big commercial one.

The Scale of Production in Rearers.

The scale of production with the use of my commercial rearer can be enlarged at will, according to the number of birds required, as several sections of rearers may be connected together with only one source of heat, and the more the scheme is enlarged the less the cost. The rearers may be worked in winter under a shed closed on three sides. Waste of heat can be provided for in the heating apparatus, which may be protected by non-conducting material. Should the winter be severe a big and light granary, or a big room, well lighted and ventilated, would answer splendidly for some thousand chickens. Naturally, under a shed where the cubic air space is unlimited, and the air is being constantly renewed, a greater number could be kept in a connected row of several sections heated from a single source, either horizontal or vertical. A commercial rearer comprises eight sections,

Advantages of My Rearer and Nest.

I would have liked to satisfy completely the curiosity of my readers by a geater enumeration of the advantages of my rearer and nest, etc., although it is rather awkward for me to praise my own inventions; but the benefit I expect to bring them

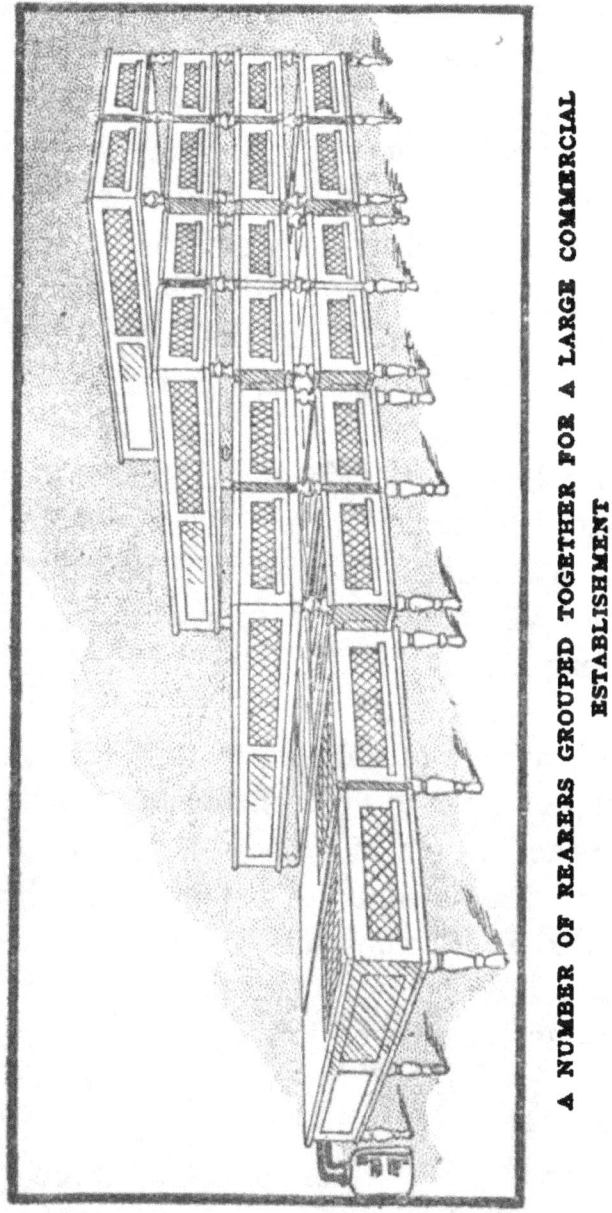

A NUMBER OF REARERS GROUPED TOGETHER FOR A LARGE COMMERCIAL ESTABLISHMENT

has encouraged me to do so. I shall be glad if I can succeed in inducing them to start according to my methods. I am convinced of the advantages

Trap-Nests to Prevent Egg Eating

they would obtain from it and from appliances which have been worked out, tested, improved, or completely invented by a practical breeder, whose time has been for years absolutely devoted to the well-being of the birds and the study of every scientific thing relating to them; whose love of birds has taught her the numerous requirements, habits and fancies of poultry; who has learned the value of time, the obligatory automatic control of a paying business, the risk of dilapidation arising from neglectful supervision, and the way to make an investment of capital pay its maximum. All these considerations have made me consider it a duty to give my readers the benefit and similar opportunities of following my experiences. Thorough knowledge of practical working helped by long practice, science, observation and continuous improvements are assets of too great a worth to be selfishly kept to oneself.

The difficulty at first was that I did not intend to deprive myself of the profit that I could legitimately expect from my patient work, huge expenses, years of practice, etc., for the sake of satisfying the curiosity of people eager to benefit gratuitously by my knowledge. I thought it would be fair that profit should be on both sides. To get over such disadvantageous publication of my appliances patents have been applied for, and my several apparatus were exhibited at the Dairy Show, on the 19th to the 23rd of October last. They obtained there a success which led me to think that my English visitors have fully appreciated them.

CHAPTER XIV

DIFFERENT METHODS OF HOUSING POULTRY

WHATEVER method of breeding, rearing, or selection be applied, very little good will be obtained from it if the birds are badly housed. Unfavourable conditions will undoubtedly check the maximum of production.

Success in poultry largely depends on small details, too often overlooked, which are so closely connected that the omission of one may greatly affect the results. The happiest and best kept hens are those of back-yard holders, to whom they become familiar pets as well as a pleasant occupation. These birds receive thousands of tit-bits that they would not obtain were they kept in greater numbers. They are not so luxuriously housed as those kept in big undertakings, but very often are better sheltered against the rain, wind and damp by walls and surrounding houses. Their life is no more natural than the life of the birds kept scientifically from which we want, nevertheless, to obtain a maximum of production under circumstances that in our presumption we have decreed to be the best for them. Are those conditions really as good for the well-being of the birds and the object in view as we think ? Are

Different Methods of Housing Poultry

the different systems of keeping birds as good for our own interests as they could be? In many cases the answer has to be in the negative. For example, in some cases birds are submitted to the terrific heat that emanates from roofs covered with tarred felt, asphalt, corrugated iron, etc.; even if those heat-absorbing materials are placed on wood or felt, does that improve their well-being? It is a bad covering, and the only way to modify the bad effects of such roofs is to paint them white in order to repel the rays of the sun. Why it has not been applied to poultry houses—has never been thought of—is a proof among thousands of the indifference with which poultry is kept. Give the birds a house and feed them, is the haphazard method too much applied, and against which I shall never be tired of struggling. Any one who possesses white garments knows the cooling effect of white. Its application will bring to my little friends the relief so much wanted in close, low little poultry houses as well as in big, high ones. Whitewashing should be avoided as it is not lasting, paint costs very little, and everybody knows how to apply it.

Defects in Poultry Houses.

If we consider the construction of poultry houses we shall not be long in finding the defects of many of them. There are, of course, in England many reliable makers of poultry appliances whose work is absolutely above criticism so far as quality, dryness of wood, easiness of erection, perfection of labour are concerned. Unfortunately not all of these constructions are perfect as regards the comfort of the birds. The greater number are imitated from American ones, partly on account of an inexplicable boom for everything of American origin. The conditions of climate being different here it is erroneous to think that the American models can be used under any conditions and without improvements, as if they were standard appliances. The birds can advantageously be kept in better conditions. The best thing that has come from America is the popularising of the scratching sheds attached to the poultry houses.

I have visited numerous utility poultry farms, big and little, in England, and have heard from their owners that they had several times experienced some drawbacks with their American poultry houses, and that they had had to alter their construction. Some of these houses have shutters meant to be open or shut according to the atmosphere. After long experience in housing poultry I do not like and would not accept appliances that have to be worked according to changeable conditions of weather, in which canvas or shutters have to be opened or closed according to the judgment or thoughtfulness of assistants, often faulty, and in which accordingly the poor birds have to take their chance. All of them do not possess, even if they are healthy birds, the same constitution and strength, therefore a single night of exposure coming from this haphazard method can affect a bird and start a cold in the house. Irregular management cannot bring regular production. Accordingly poultry houses have to be devised to keep the birds in good condition under all circumstances.

Different Methods of Housing Poultry

The Moment for Poultry Keeping.

Never was the moment so propitious as now for starting utility poultry keeping. France and Belgium will hardly be able to supply their own need for a long time. Never was sensible advice so much wanted, but the novices must be very careful not to start badly if they want to obtain profit from their undertaking. At the moment poultry keeping is advocated to small people with a view to inducing them to keep fowls, and to those who are already keeping a few, to keep more. Instructions for building poultry houses themselves and keeping hens cheaply are given.

A Simple Method of Poultry Keeping.

My own advice will be still simpler and cheaper than those already given. I would say to small holders, simply construct a lean-to or gabled roofed shed, rather high. Let the birds perch at the top, underneath the roof, and they will find at once the best place for themselves. They will then be well sheltered from wind. Surround the shed with wire netting, and hang at the bottom a width of tarred felt round the outside by means of small hooks to shelter them from the wind. Nothing can be cheaper, and the birds will have plenty of fresh air and no draught. From the first day they will adopt their own perch, underneath which it is easy to place a dropping board if desired. Feed them well and the birds will be healthy, hardy, and would, if selected for laying, under this system be absolutely profitable, but take care to begin this system in the spring. Nothing is worse than a bad house, and I know that no house at all is better than a bad one. I also know the splendid effect of continuously renewed fresh air for them. During the last five years the birds I kept for my own consumption had never been sheltered otherwise, and that in a very low and damp climate, and they have been perfect in every way. They must have a shed as high as possible, and they will never be affected by the heavy dampness floating some feet above ground which invades and remains stagnant in badly ventilated buildings.

Egg Production as a Business.

Of course, if a business is being made of egg production buildings are necessary, and must be differently constructed to the ordinary poultry house actually in use. Those are built, as a rule, in a long continuous row of one or two stories. Thousands of birds are kept in such houses in groups of 100 to 250, and sometimes 500 together. They generally never go out, as they are kept under the so-called intensive system, but some breeders are obliged to let them out from time to time, on account of the unsatisfactory condition of the birds. Some other breeders have adopted the semi-intensive system, which means that the birds are kept inside on bad days and are allowed to go out in the yard in fine weather. There is much to say about these systems, but for the moment we are considering nothing else but the housing, and the effect of it on the birds. Every time I visited big plants of this kind I had to pity the poor inhabitants, and more so

Different Methods of Housing Poultry

those living in the second floor, as the heat coming from the roof over them made them pant, increasing their misery.

Bad Systems.

In one such place the keepers were not very eager to take me to the second floor, but I managed all the same to go up. I soon understood their objecting to my visit when I saw the miserable-looking appearance of the birds sleeping under the lower part, and the terrific heat from the sloping roof. Those houses, nearly everywhere, are well kept as far as dropping boards are concerned, but they do not fulfil the hygienic conditions for birds kept in semi or total confinement. These intensive and semi-intensive systems have been alternately praised and criticised to me by renowned breeders living in various parts of this country. They succeed in some hands and are a failure in others, in spite of the cleverness of the breeders. The reason for this difference of opinion results in conditions seldom the same. Here it lies in the alteration of the principle to suit different climates. In one case because the birds are kept in too great a number. In another because the floor space is not sufficient for the number of birds, or on account of unsuitable food being given. Also, in a great number of cases, on account of the lack of knowledge and thorough cleanliness, obligatory when birds have to live for months in the same place. In these houses contrary to what one would expect, birds are not generally trap-nested, and two or three hens are permitted to lay in the same nest together. Nothing could be worse for the eggs or hens. In short, to enumerate all the causes of failure would require volumes.

The ideal utility poultry house has to be conceived on sound general principles, and adapt itself to all conditions without the interference of more or less intelligent people. It must avoid the use of blinds, shutters, etc., which are impracticable and subject to continuous repairs and expense. Nothing adapts itself so readily to its surroundings as poultry, provided that certain rules are observed, and very little is necessary to satisfy the birds. The inability of some people to detect the requirements of the birds in the big intensive houses has been strange to me, although it is, in a way, human that people having paid from £100 to £500 for such intensive houses should not be inclined to find defects in them. They realise that something is defective, and naturally lay the blame on the birds.

In such houses one would think that birds, not being allowed to run about nor to get wet or exposed to the wind, would never be subject to cold; but that is not the case, as I heard a lot of birds sneezing and saw some serious cases of colds. My remarks were agreed with, that some cases of roup break out from time to time, and it was added, " We wonder why, with birds so well protected." As I was not willing to argue and hurt the feelings of the proprietor, who, having paid a good price, was led to think that he had obtained the best, I kept silent. In spite of all this, such businesses pay, but how differently would they pay if the birds were kept in more suitable buildings. Disease would be then avoided, as well as disinfection of the house; the changing and burning of the litter will no more be necessary.

Avoiding Roup and Other Troubles.

No cases of roup or other ailments coming from cold should arise in a poultry house so constructed as to guarantee the birds from damp and cold winds, as well as providing them with a continuous renewal of fresh air. To go into the experiments which have taught me the best way to house poultry profitably would require too much explanation. It is sufficient to say that all the mistakes that are generally made by the novice were made by me. I had no more intelligence for poultry breeding than the person most ignorant on the subject. I had only the great advantage of not being handicapped by lack of money, so that I could try different systems for years, till I found a system which answered so splendidly in the health and production of the birds that I discarded all other buildings and had all my poultry houses built on the same principle. For nine years I kept birds of different breeds, some very rustic, others very delicate, and I have never heard the slightest sneezing since they lived in these houses. I also obtained from my birds a regular and large number of eggs. This was the reward allowed to me for the terrible disappointments, worries, illnesses, and hard work attached to poultry keeping in inexpert hands. Thanks to my love of the birds and my persevering character I was never disheartened, and continued experimenting till I was able to point out to other people what, to my mind, was good or bad.

Extremes of Temperature

Success of my Methods.

My first article " Poulaillers ouverts en toutes Saisons " would have brought me thousands of visitors if I had not stopped this invasion the first year after the visits of some hundreds of people. My time was too precious to be spent in showing my poultry yards. No visitor would at first believe that birds living under this system could lay so splendidly. That very year the winter had been so severe that eggs just laid were in some seconds burst open by the frost in the nests. The condition of the birds with their beautiful red crests and tight shining plumage was the wonder of everyone, as was the laying record. The consequence of those visits was that my system was followed as closely as possible by people grateful enough to call themselves my pupils, and who obtained the same fine results. I had the satisfaction of seeing the slightest detail found in my houses followed religiously by those pupils of mine.

In my poultry houses no stagnation of air or damp is allowed, which is the case when the birds in the big intensive or semi-intensive constructions roost opposite the wire netting front, exposed to wind and mist. Underneath the lowest part of the sloping roof they are badly sheltered against a too direct exposure. They are overheated in summer by the warm roof over their heads, and in autumn or winter exposed to the prejudicial effect of the mist, which penetrates their feathers and keeps them damp most part of the day. The air is sometimes permitted to escape through eaves over the hen's heads, carrying

Different Methods of Housing Poultry

away with it the beneficial heat generated by them, and which should have helped them to rid themselves of the moisture that clings to their feathers. As far as the mist itself is concerned, being heavier than the air, it remains in the house and lies on the litter. When the birds awake they jump on to the damp litter, from which neither the heavy stagnant gases nor the natural moisture of the excreta have been removed, and to which an additional dampness has been added in the course of the night. How can one be astonished at the colds that affect certain birds? What a demonstration of their enduring qualities, to see birds living and yielding a profit under such unhealthy and contrary conditions, especially when one knows that the litter remains for months in the building without being renewed. Added to this, that the litter is generally shaken or cleaned of the excreta when it is done in the day time underneath the noses of the birds, who are thus obliged to swallow the dust and breathe the emanations from it, and you have every reason to be full of wonder at their robustness.

I have kept my birds for nine years under the intensive system; six months every year—during winter production—in roosting places permitting them to breathe a continuous supply of fresh air, without submitting the birds to the least draught. The litter was shaken after the birds had gone to roost and allowed to dry all night. Not the slightest smell could be detected in the poultry houses, and although several breeds, such as Polish and Hamburg, are easily affected by cold and dampness, these conditions suited them all. I was at the time

living at Liége, where the conformation of my
ground obliged me to have the houses placed in a
very exposed position and in a damp, draughty
place, but I never had to regret the steps I had taken
after years of irresolution and tests. The birds were
never kept more than fifty together, but my poultry
houses were so constructed that each one could be
attached to another so as to make a continuous row,
and made into one building, so that the plant could be
extended to 160 ft. or 200 ft. in length by additions,
as well as being constructed in stories, keeping,
thanks to the ventilation and several other features,
birds in the top story in the same fresh and healthy
conditions as those on the ground floor. The
present difficulty of labour has made it impossible
for me to construct such a house for the Dairy Show,
but a reduced model has been exhibited with my
rearer, nests, and hen's harness at Messrs. Spratt's
stand. It has been a practical illustration how it
was possible to give birds open-air conditions,
shelter, facility of selection and attention, better
than pages of description.

CHAPTER XV

THE DIFFERENT SYSTEMS OF KEEPING POULTRY

THE Intensive System on a great scale is not always as well understood as it ought to be. It is the ideal method for the commercial production of eggs; but up to now, the most essential points of hygiene have not been sufficiently applied to say it has been a complete success. It is not a new system; in fact, long ago before the boom of the American Intensive Poultry Houses it was known in every country of the world. Wherever birds have been kept in poultry houses, without being allowed to go out, the Intensive System was practised. It is the best way of obtaining the maximum of production, although I have found that which answered the best was a combination of the Free Range System applied at a different period of the year with the Intensive keeping; for example, my future layers were reared till six weeks old in confinement, but with plenty of fresh air, then allowed full liberty during their youth till they were on the point of laying. This is easily detected, because the age at which a bird of a certain breed lays is always about the same; a light breed lays generally at about five to six months old, a heavy one at about six and a half to seven

and a half months. One month before the time for
laying is expected my birds were progressively kept
in confinement till the beginning of March, at which
time the young pullets as well as the older birds
were progressively permitted complete free range
till moulting time. These five months of complete
liberty, the change of diet, the access to the grass,
the search for the thousand little eggs of insects
and worms, their separation from the cocks—when
breeding is no more necessary—the good effects of
spring and summer, all these conditions together
kept them perfectly fit and in a fine state of health
seldom met with in birds kept in the ordinary Intensive System. No deterioration of stamina can
take place if birds are so kept when they live in
sweet and pure surroundings with plenty of air
and perfect ventilation of the litter and sufficient
exercise. One important point to take into account
is that the change from liberty to confinement, as
well as change of diet, must be brought in gradually.
Hens are creatures of habit and should be dealt with
carefully wherever production is concerned. A
check in laying means fewer birds to rear together
and to sell with the same labour and nearly the same
expense of light, fire, litter, attendance, etc., than
if they were more numerous.

No Feather-Pullers.

My birds never need when kept in this way,
to be allowed to go out to pick up strength,
as they never lose it, thanks to the activity and
amusement they find in their houses. Busy birds
have this analogy with human beings that they

Different Systems of Keeping Poultry

have no time to be vicious. There is no pulling out of feathers in my poultry houses! My birds have never been too closely in-bred, and the cocks when not in service are never kept in the proximity of the hens. They are bred in full liberty and in other surroundings, on a different diet to that of their mothers and sisters. An opposite mode of living is known to help to modify the dangerous effect of in-breeding. Cocks until eighteen months or two years of age never approach the hens reserved for breeding purposes; should cockerels be required for mating they should be fully matured and developed before being utilised. In the months of September and October the young would-be laying birds which come from the best layers are picked out of the yards and placed in their definite mating pen, they are never stimulated by special laying food but judiciously fed. Then, when they have laid their first batch of eggs they are mated, if the production is meant for the rearing of table fowls, with a well-developed young cockerel.

Strangely enough, it has been found by our fatteners that chickens coming from young stock are not only the finest and most succulent fowls, but that they fatten quicker. In France and Belgium they insist on chickens coming from young stock for this purpose. The best two-year hens from which a laying or breeding strain is expected are selected, and kept alone until ready to lay, then mated with two- or three-year-old cocks, but when the production is meant for consumption no cocks at all are allowed with the hens. For whatever purpose birds are utilised, in the period of production the food has to be differently balanced than when the birds are at rest or moulting; it has to be richer in egg-forming material and reinforced with essential nutrients and mineral salts. From birds so fed most robust chickens may be expected.

Hens Kept in Rooms.

The Intensive System does not require much land, practically no land at all unless it is carried out on a big scale. I have seen in different places birds kept in a granary or in rooms doing perfectly well. Of course, when birds are kept in this way it is advisable to get rid of them after the intense production is over, thus avoiding the keeping of the birds during summer and moulting time. Other birds should be bought for the following season, if it is not found possible to send them into the open during summer time. Birds so kept are not desirable for the building up of the future laying stock, such young stock being liable to breed delicate offspring. The difference in price between the birds sold in the spring after winter production and the birds bought in the autumn for winter laying would be balanced by the economy in the cost of feeding and by the production in winter. There is no doubt that money is to be made in this way of keeping poultry, especially in small undertaking, when the care of it can be entrusted to the housewife and where the birds can be housed or kept without investment of capital. I read some days ago that thirteen shillings and some pence could be won on each hen; perhaps it might under certain conditions; if no depreciation of material, no labour, no

Different Systems of Keeping Poultry

expenses of building, etc., is necessary, but such profit depends too much on special aptitude, knowledge, hard work, etc., to be accepted as easily obtainable. From five to eight shillings a bird may be expected if people are careful enough to buy birds from good stock, give them suitable accommodations, keep them clean, and feed them judiciously.

The Semi-Intensive System.

The Semi-Intensive System consists in letting the birds out when the weather permits. Here again judgment is necessary and cannot be relied upon in commercial business; whenever, as already pointed out, the breeder and the birds have to rely on other people's services and thoughtfulness they are already handicapped. A day that has begun splendidly may end injuriously in cutting wind, cold and damp, and be fatal to the birds. Birds going in and out occasionally are more sensitive to the weather than if they were living in complete liberty. Apart from that, the litter of the house on a rainy day gets wet, retains the humidity and keeps the birds' feet damp all day long—as well as in the night if the litter is not properly ventilated. This mixed system with only occasional freedom exposes the birds to the risk of illness. It has neither the practical advantage of the Intensive nor of the Open-air methods. Some birds in the Intensive system are too often obliged to be let out on the grass, having neither possibility of exercise in the poultry houses used in this method, nor good hygienic surroundings; their poor condition shows itself in their pale faces and combs.

The " Free Range " System.

The poor birds would probably die, at all events would be useless as productors, if they were not allowed to go out to pick up strength. Can these be called Intensive or Semi-Intensive methods? Half-measures are never good, and birds, should be kept in completely, but in healthy surroundings, or they should have their entire liberty. The Free-Range System has advantages and disadvantages. I have already said that my breeding stock enjoy it. The only improvement I made was to spare the first days of my baby-chicks, for I saw no necessity to sacrifice part of them in exposing the chicks as soon as born to the inclemency of the weather; but so soon as they are six weeks old they are allowed to run and get wet as much as they like. If some of them do not bear cold or dampness, as they are bred for the making of strong stock, if they are too delicate, it is much better that they should not live; but this seldom happens, and I find them more vigorous than brought up in the other way. One of the disadvantages of the Free-Range System is that it cannot give a maximum of profit nor regularity in the production of the eggs, owing to the sensitiveness of the birds to atmospheric conditions. Every poultry keeper knows that some days of rain, snow or cutting winds bring generally a cessation of laying. In a commercial undertaking it would not be possible to rely on free-range birds; also some of them would certainly lay astray.

Different Systems of Keeping Poultry

Eggs are not only required for consumption and sittings, but for the intensive production of table birds. Its advantage lies in cheapness of erection and general outlay.

Birds in Yards.

There exists another method of keeping birds, which is applied nearly everywhere in England. Birds are kept in yards, the poultry house is enclosed by fencing, they may go in and out as they like, their yards allowing them more or less space. This system has the advantage of the Free-Range that no eggs can be lost; but it has this disadvantage, that birds do not enjoy the same exercise and do not find part of their feeding as they would under the Free-Range method. It exposes them to the same vicissitudes of atmosphere, and is rather costly in land, fencing, labour, and supervision, without having the advantage of intensive keeping. In many places houses, fences and birds are moved on to the land with a view to give them a sweet soil. Were I to keep birds in this way, and find it necessary to change the land, I would rather have double or triple yards and allow the birds alternatively on them than to have this tiresome business of moving them altogether. When all this labour and supervision are taken into account it costs yearly more than the interest of the capital required for a scientific installation, and the profits are less.

Returning lately from the country I saw poultry yards of this kind with the soil covered with snow. No birds were to be seen; they had had to take refuge in their little houses. Poultry hate snow and object very much to go out in it. Of course the poor birds in their tiny houses are moping in a corner, with no mind to lay. The same thing happens in heavy, rainy weather.

Incubation in October.

We obtain in Belgium and France eggs for incubation purposes in October, and regulate the birth of our birds in such a way that they lay when we want them to do so. Nothing has interfered to prevent their development in their young days. Consequently, our birds begin to lay at the moment we have determined; apart from the natural rest which they enjoy between two batches of eggs, we can fairly rely on their laying, because we do not permit our birds to shiver on the grass at the moment when eggs are for us so precious. The first chickens obtained in England are, setting apart some few commercial establishments, reared from their first days on the grass. Fortunately, English chickens appear on the market only in the month of March, at the moment we generally cease to incubate. A profitable winter production would be totally impossible under the English method. We begin to sell at the end of December, and these three extra months of production make all the difference in profit. Unless artificial methods in rearing and keeping are applied, it is totally impossible to regulate the production. The Free-Range or penned-bird systems breed a sound, healthy strain, no doubt; it is the survival of the fittest; but, as I have already shown, it is a wasteful system in life, suitable to people who

Different Systems of Keeping Poultry

cannot afford to invest the capital necessary for the setting up of an intensive poultry establishment. It is always possible to get from birds, if they are cleverly handled, a fair profit; but a maximum of production will never be obtained, neither in eggs nor in chickens, in this way. It will appear strange to my readers to read that I prefer to breed from birds kept in confinement, provided they have been reared till laying time in free-range. It seems paradoxical and contrary to what is generally thought, but numerous experiences have proved to me that the fertility of the eggs and the vitality of the chickens born from captive parents are far superior to the fertility and vitality of eggs coming from birds partially or completely free. One of the reasons lies in the birds having never been subjected to changeable conditions of weather. How often the most robust birds are to be seen in bad weather shivering and moping; when the weather is too severe, even the most hardy of them are undoubtedly badly affected by it, and the fertility of the eggs is lessened, as well as the vitality and hatchability of the chicks. Many times I recorded the eggs fecundated in those periods of depression, and always found the chicks more delicate and more difficult to rear when coming out of such eggs. The momentary dispositions of the parents have a reflective effect on their offspring. In my way of keeping birds I never had the slightest difference in the fertility and vitality, even when they were conceived and reared in a most severe winter. I attribute this considerable advantage to the happy condition in which birds are kept in my ways and partly to judicious feeding, better controlled when birds are under the complete care of intelligent breeders.

Strong Mother Birds.

Science tells us that breeding should begin in the mother; it means, that birds intended for breeding purposes should not only be kept and fed accordingly, but that the mother of our future stock ought never to have been delicate. She has to be not only constitutionally strong, but must possess a reserve of all the elements that will enable her to produce, without detrimental effect to herself, robust chickens. These elements ought to be daily renewed if we want her to bear the strain of intensive production; unfortunately it is not always the case. Very often I have seen the ration curtailed in moulting time under the pretext that the hens do not lay and consequently do not need so much food; the same neglect occurred with cocks when not in service; the consequences were that the poor birds had to live on themselves and weaken their constitution just at the moment it would have been wise to help them to renew their coat of feathers. This lack of foresight has wrought havoc among poultry keepers.

In moulting time not only suitable food but green stuff should be liberally given, the food ought to be cooling but plentiful, over-heating mash and over-heating grains avoided, an addition of nourishing phosphates and mineral and vegetable salts given to them—it will help the birds wonderfully not only towards feathering, but towards laying, because it brings into their constitution phosphorus matter, iron, etc.—whilst the ration should be higher in protein and fat; with birds thus treated, a great part of them, even in moulting time, do not cease to lay. It is a great error to think that these mineral and vegetable salts can be obtained from vegetables, because vegetables contain firstly five or six times their weight of water, and secondly are too bulky to permit a great absorption of them; the birds whose crops are limited can only extract an infinitesimal portion of salt included in them.

The " Bulkiness of Vegetables."

According to Hutchinson, " The bulkiness of vegetables stimulates by its mechanical action the movement of the intestines, and the contents of the latter are thus so hurried on that not enough time is allowed for the complete absorption of nutrients to take place. Unless vegetables are still young and very tender a part of them is converted by the action of bacteria into marsh gas and consequently lost." If we add to this assertion that a great part of the vegetable bulk is cellulose and is imperfectly attacked by the digestive apparatus, we realise that vegetables are not very advantageous to birds so far as mineral salts are concerned. Therefore in an intensive production there must be a renewal daily of those constituents, without which no egg can be built. Birds living a natural life and not being forced to produce generally find themselves suitable food, containing sufficient of these salts, for the reduced amount of eggs they lay. Wherever intensive production is practised—we have seen the fact in cattle and other animals—concentrated food has had

"Nous avons changé tout cela"

to be added to the diet. Animals are wonderful transmutors of food, and hens particularly turn all food to the best account. There is a saying in France, " La poule pond par le bec," which means that if you give her what she requires she will make you a return of an egg. The hatchability of the eggs depends not only on the condition of the hen, but upon the food; a weak chick will die in shell. An egg is a chicken which has not yet come to birth : it contains blood, muscles, flesh, nervous system, feathers, bone, horny matters, etc., all substances made from mineral matters, protein and fats. This fact, always kept in mind illustrates the importance of food, especially where an intensive production is sought for. Some time ago a striking article appeared in *The Reliable American Poultry Journal* concerning hens that individually laid ten eggs a week. Without going so far, I want to point out that in Nature, at the time of the year birds are meant to lay, the days are very long and the nights very short; the birds eat from morning till night at the moment when they are in full production. We have changed all that and have trained the bird to give us eggs at the time of the year suitable to us. If the birds have to give us in quantity what we expect from them, it seems only logical to place them in the same condition as when living naturally. In winter the nights are very long and the days very short, conditions are no longer favourable for production, consequently unfavourable for extra production. Our thoughtfulness has to intervene. We must artificially provide the birds with longer days, compose our rations in such a way that they should

Different Systems of Keeping Poultry

find in them not only what they find in full liberty, but the heating elements that help them to be indifferent to the degree of atmosphere.

Analysis of an Egg.

The composition of an egg is as follows:

	Water	Proteid	Fat	Other Non-nitrogenous Matter	Mineral Matter
White	85·7	12·6	0·25	0·13	0·59
Yolk	50·9	16·2	31·75	—	1·09

The importance of water is pointed out in the above enumeration. Pure water is necessary. Contrarily to what many people think, rain water is not suitable for birds; it contains the impurities of the atmosphere, and an absence of lime which is so necessary for the composition of the shell. To understand better the requirements of our birds let us see the mineral matters enclosed in the egg:

	Yolk	White
Potassium	9·29	31·41
Sodium	5·87	31·57
Lime	13·04	2·78
Magnesium	2·13	2·79
Oxide of Iron	1·65	0·57
Phosphoric Acid . . .	65·46	4·41
Sulphuric Acid	—	22·1
Fluorine	0·86	1·06
Chlorine	1·95	28·8

Comparison of Analyses.

Years ago I had analyses made from eggs coming from Free-Range fowls and from fowls kept intensively but well fed. The result was very similar in both cases, but later on, having invented my phosphates, the better result obtained in incubation of eggs was so noticeable that I wanted to know what enrichment had taken place in the eggs after using my phosphates. The yolk was of a rich yellow, the egg less watery, more condensed, therefore more nutritious. No doubt was left to me after the analysis of the importance of an increase in phosphates and mineral salts. No more deaths in shell, and the most vigorous and easily reared chickens. Roughly speaking and without entering into the important subject of feeding, which would require another volume, I used to balance my ration in extra production as follows : for one part of protein I allowed four parts of carbohydrates and fat to which a proportion of my phosphates was added, and I had the satisfaction of seeing my birds in the best condition of health and of production. Of course an abundance of green food was always given, grass or sprouted cereals and seed. This green food was minced and added to the morning wet mash, this latter being more a vehicle for conveying green food and concentrated phosphates than for the sake of giving wet mash. The sprouted tender grains are obtained by treating cereals and seeds in a special appliance which enables me to get enough green to feed hundreds of birds all the year round. Some beetroots are also hung sufficiently high in the pens to oblige the birds to jump to get at it ; it provides them with exercise and amusement.

The mash referred to above is given at ten o'clock ; when the birds get up they find in their litter cracked grains for which they have to scratch. As the spreading of these grains in the litter throughout the night would attract rats and mice, a special little device when the first hen steps on the stairs

Different Systems of Keeping Poultry

releases a tin-full of grain which scatters the
tiny bits about the litter; this keeps the bird
busy until the wet mash comes. At one o'clock
they are allowed to get to the dry mash given in
linen hoppers, from which they eat as much as they
like; then at about three in winter birds go to roost
without anything else. At seven o'clock my
poultry houses are lighted and the birds may eat as
much as they like of whole grain. The dry mash of
the afternoon has completed the wet mash of the
morning; it has given them, in a concentrated
form, all the highly digestible nutrients necessary
for egg production. At seven o'clock the crops of
the birds are already empty. They refill themselves,
this time with the harder matter of the grains. All
the day the birds jump, scratch, cackle, and quarrel
to their hearts' content, spare feeding has kept them
active. Nothing is so indicative of good health.

Where there is no possibility for lighting poultry
houses there should always be a way to light the
roosting place in order to allow the birds to take
their last food. After one hour the light should be
lowered in order that the birds should return to
roost, then extinguished. The extra food gives them
enough nourishment to wait for dawn. No mixed
grain is given to the birds, but only one sort at the
time in order to avoid their cramming themselves
with certain favourite grains; also because I want
them to get alternatively the different substances
of the different grains. Mixed grains thinly cracked
are given only in the morning. In severe weather,
some cracked maize is allowed. Indian corn or
maize is given very sparingly. This way of feeding

and keeping birds has given me such beautiful results that I would not apply other ways either for eggs for consumption, or for the production of table fowls.

Poultry keeping no doubt will become in the future a great industry in England. One hears sometimes that no money is to be made in poultry. It is a ridiculous allegation and as stupid as the statement that it is possible to make a profit of over thirteen shillings a bird. That there is money in eggs and table poultry is sufficiently demonstrated by the imports into this country. Eggs come from Russia, Bulgaria, Denmark, Italy, Belgium, France, etc. People do not generally send goods if they obtain no profit from them. Indeed, the profit must offer a good margin, to permit the exporters to bear the expense of boxes, packing, and shipping the eggs. If there has been a deficiency in eggs in England, there are many reasons for it. One of them can be attributed to an unsuitable climate, against which birds have consequently to be guarded. I have been very much astonished to find that poultry experts who have travelled extensively have not realised that as the atmospheric conditions of England are different to the Continent, the keeping of birds could not be the same. Different conditions require different management. The laying competitions in this country have done much for the building of some laying strains bred by clever poultry people—they are too few to produce but a limited number of real good layers; they constitute up to now the exception. The majority of breeders do not succeed. Whatever push will be given to poultry keeping will be useless, unless some changes occur. Not only are people slow to take to new methods, but the poultry press does not show eagerness to enlighten its readers when some new methods are opposed to theirs. This is unfair and prejudicial to progress. Truth, however, bears in itself enough strength to come out, and I am certain that poultry will undergo a change for the better. In fact, it has already begun to do so. Better classes of birds will come on to the market. Eggs, with better feeding will be more nutritious, and chickens coming from them more vitalised. The inferiority of table fowls will slowly disappear; this would be more satisfying to both buyers and producers. The time for poultry has come, and although prices are higher than they were, poultry started on a paying basis on a small or great scale will pay according to the number of birds kept and their management. My personal experience has been that the price of food is not an objection to a good market. Better quality commands a better price, and compensates for the rise in food. This fact has been illustrated in the past winter by the high price obtained for eggs. The lack of foresight and fear of food prices have led many poultry keepers to get rid of their birds, and they have lived to regret it. In a short time, if I am not mistaken, eggs and table-production will industrialise themselves. Some large plants will be built which will prove that well-started and well-managed establishments will bring profit. They will be erected by private people. The Government

Poultry Keeping a Duty

does not appear to me eager to show the lead. It advises every landholder to keep poultry for the good of the land and the production of eggs or chickens—but that it will be better in the future is rather doubtful so long as the old methods are kept. People want to see the result of new methods before going into them. I trust to the eager and patriotic feeling of some rich gentlemen, clever enough to see what there is in poultry and the good they can be to their country and country folk, rather than to any change coming from official quarters for the erection and spreading of scientific establishments and appliances.

Everybody should in the future keep poultry. It has become a National want—no land is necessary for keeping a few birds. The manure from fowls kept by small holders is of considerable financial importance to the country. It finds a ready sale for agricultural and horticultural purposes, for grapes, tomatoes grown at home, etc. The manure of birds kept intensively is better, in fact, and stronger than manure coming from other fowls.

I hope in the near future to be able to demonstrate in a model establishment all the management necessary for big and small holders; all the branches of poultry keeping beginning from the rearing of stock, to the selection of eggs, the production of eggs for consumption, for sittings and rearings, the trade of newly-born chicks, the production of *Petit Poussin*, of chickens for the table, etc., and to make people realise that there is no better profit, considering the amount of outlay

Different Systems of Keeping Poultry

and the price of stock, than poultry keeping, provided one has acquired sufficient knowledge of it, has some business ability, and taste for the work undertaken, without which failure is certain.

Before closing my book, I owe thanks to my readers for the numerous proofs of appreciation and sympathy I have received from them during the appearance of my articles in *Country Life*, and also at the Dairy Show, where my appliances have met with such success. It has shown me that my system and advice have already been found useful and effective in the country—which is for me a dear refuge deeply appreciated. I could ask no better reward.

Printed by Hazell, Watson & Viney, Ld., London and Aylesbury.

SIDE VIEW OF THE REARER.

This illustration discloses the inside of the outer runs, and the little stairs in the centre show the access to the playing grounds. In the lowest section the playing ground is exposed to view.

IV. TWO SECTIONS OF THE REARER.

Showing the Shutters closed in the sleeping compartment. Note the ventilating
tube which can be used outside.

III. TWO SECTIONS OF THE REARER.

This illustration shows the covering for use out of doors and the runs and heated chambers opened for cleaning.

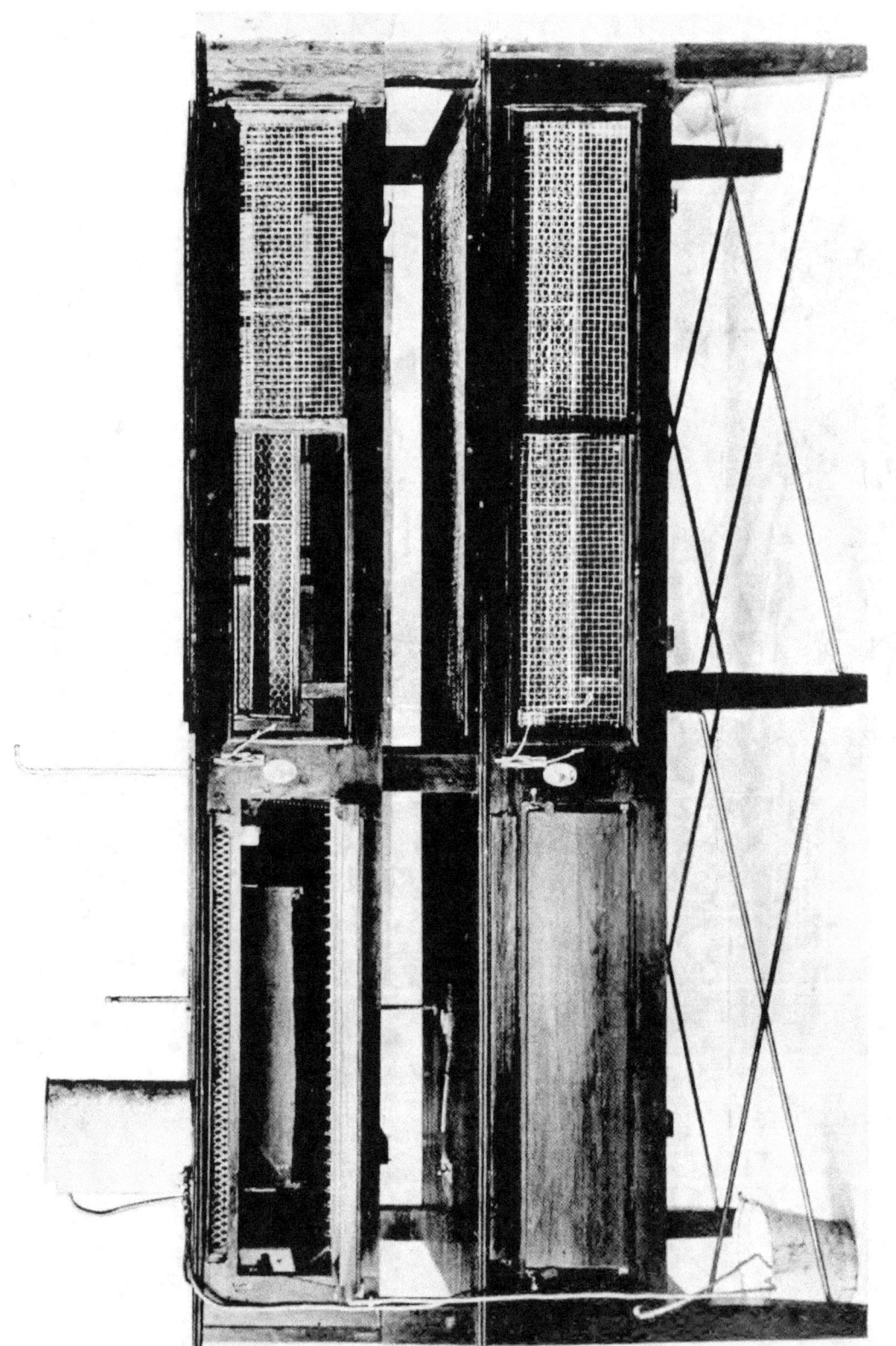

II. TWO SECTIONS OF THE REARER.

This illustration shows the thermometer mechanism for lifting the valance in the sleeping (or heated) chamber, the ventilating tube, drinking vessels, water tank, pipe and waste.

I. TWO SECTIONS OF THE REARER.

In the heated compartments the shutters are shown open and in the open yards the playing grounds are seen. A wide air space is provided between each floor in order to isolate the birds. Wire netting is employed to permit of the freest air circulation.

II. INSIDE THE HOUSE.

Everything closed, with the exception of one trap nest which is shown open.

I. INSIDE THE HOUSE.

This illustration shows part of the roosting place open, as all the four doors should be during the day. The little trap door is also shown open, disclosing how the hens are released.

ANOTHER BACK VIEW OF THE HOUSE.

This illustration shows the perches suspended from the roof by metallic rings. (In this and the preceding illustration the back wall has been removed for the purposes of the photograph.)

BACK VIEW OF THE HOUSE.

This illustration shows the double roof and the corridor, permitting the freest circulation of air. Here the bottom trap is seen open and shows how the litter is ventilated after it has been cleansed. The trap closes automatically when the hen steps out of the roosting place.